DESERT TOURISM

DESERT TOURISM

TRACING THE FRAGILE EDGES OF DEVELOPMENT

EDITED BY **VIRGINIE PICON-LEFEBVRE** WITH **AZIZA CHAOUNI**

THE AGA KHAN PROGRAM AT THE HARVARD UNIVERSITY GRADUATE SCHOOL OF DESIGN

CAMBRIDGE, MASSACHUSETTS

ISBN 978-1-934510-18-6

Book and cover design by Wilcox Design
Printed and bound by Kirkwood Printing, Wilmington, MA
Distributed by Harvard University Press

Cover image: Quartzsite, Arizona, courtesy Robert Sumrell and Kazys Varnelis

The Harvard Design School is a leading center for education, information, and technical expertise on the built environment. Its departments of Architecture, Landscape Architecture, and Urban Planning and Design offer masters and doctoral degree programs and provide the foundation for its Advanced Studies and Executive Education programs.

CONTENTS

VIRGINIE PICON-LEFEBVRE

INTRODUCTION: DESERT TOURISM THEN AND NOW

The goal of this book is not to take an exhaustive approach to the topic but rather to attempt to introduce it from a multidisciplinary perspective. These essays aim to untangle some of the narratives that have contributed to shaping desert tourism as a contemporary phenomenon. Two questions define the project: How might it be possible to avoid transforming every place into a theme park and instead create valuable and successful tourist venues without turning them into artifacts? And how could desert tourism foster the sustainable development of territories?

It is a great challenge for architects, urban designers, and everyone else interested in the subject to improve or create places that could be interesting for tourists but do not ignore the needs and aspirations of the inhabitants as well as the local ecology. We know that rising numbers of tourists often interfere with the interests of the inhabitants—destroying sites, natural resources, landscapes. It is evident that the construction of out-of-scale hotels, malls, and souvenir shops is a serious threat. One might argue that successful tourism anywhere destroys the local identity and the existing qualities of sites, replacing them with standardized, banal buildings and services. Yet because mobility, networks, globalization of the economy, and the development of travel characterize our modern developed societies, tourism also brings welcomed infrastructure, hotels, restaurants, and even arts and crafts that improve local economies and services. Without tourism, many places, especially in developing countries, could not survive.

Tourism is changing from what it was less than twenty years ago, and desert tourism is among the new trends. If the transformation of cities and other places follows a constant evolution, recent developments in tourism show that there is growing competition among destinations, with some experiencing major difficulties in attracting a new generation of tourists. For example, the Canary Islands, once a prime destination for baby boomers, must now find ways to attract younger travelers. At the same time, a new generation of tourists are emerging, looking for more authentic experiences and not interested in traditional mass tourism destinations. The growth in ecotourism is one reflection of this trend. The internet has also changed the travel market, which is increasingly defined by individualized tastes. Companies and travel agents are offering different types of destinations and experiences from those provided a few years ago.

In this context, deserts are considered the last frontier. They have lost their stigma as inhospitable, inaccessible places and are becoming an ever more popular tourist destination, with an image of authenticity. The growth of this tourism, however, is raising particular challenges for the desert and its oases, jeopardizing fragile ecosystems and straining scarce resources. Paradoxically, the increasing popularity of desert tourism is undermining the very basis of its allure. In developing countries the consequences are often even more dire, as local populations, gathered around oases, live in difficult conditions with scarce resources and insufficient infrastructure, rarely benefiting from tourism's positive economic effects.

The complexity of tourism demands a multidisciplinary approach, especially when one deals with the subject of tourism in the desert. Yet most existing literature takes the strict point of view of one discipline—most often a sociological or historical perspective. The architecture of tourism is rarely explored. The goal of this volume is to engage writers from diverse disciplines, including architects, engineers, artists, and social workers.

Desert tourism has evolved from the confluence of multiple factors and movements (colonialism, Orientalism, romanticism) as well as the desert's depiction in literature, film, sociology, anthropology, and architecture. Here we discuss issues raised by desert tourism and architecture along three lines of questioning. From a historical perspective, we will confront, through the essays of Susan Miller, Claude Prelorenzo, Neil Levine, and Aziza Chaouni, how the construction of the *imaginaire* of the desert by guides, movie directors, writers, and architects invented a narration of desert travel. Then, Robert Sumrell and Kazys Varnelis, Alessandra Ponte, and I will follow the traces of previous conflicts and show how the desert was transformed from war to peace as a touristic destination. Finally, on current conditions, Vincent Battesti addresses oases and Chris Johnson describes the Jordan desert as a sanctuary. Gini Lee concludes the volume by describing how the Australian desert inspired her production of resonant place-specific artworks.

SELECTED BIBLIOGRAPHY

Brendon, P. *Thomas Cook: 150 Years of Popular Tourism.* London: Secker and Warburg, 1991.

Britton, S. "Tourism, Capital, and Place: Towards a Critical Geography of Tourism."
Environment and Planning D: Society and Space 9: 451–478. 1991.

Eco, Umberto. *Faith in Fakes: Travels in Hyperreality.* London: Minerva, 1986.

Fine, B., and E. Leopold. *The World of Consumption.* London: Routledge, 1993.

Judd, D.R., and S. S. Fainstein. *The Tourist City.* New Haven: Yale University Press, 1999.

MacCannell, D. *The Tourist: A New Theory of the Leisure Class,* second ed. New York: Shocken, 1990 (first ed. 1976).

Page. S. *Urban Tourism.* London: Routledge, 1995.

Prelorenzo, C., and A. Picon. *L'aventure du Balnéaire.* Marseille: Editions Parenthèses, 1999.

Towner, J. "The Grand Tour: A Key Phase in the History of Tourism." *Annals of Tourism Research* 12(3): 297–333. 1985.

Urry, J. *The Tourist Gaze: Leisure and Travel in Contemporary Societies.* London: Sage, 1990.

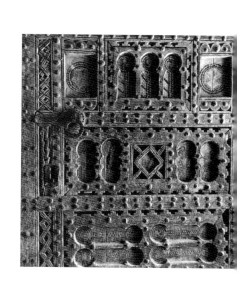

SUSAN GILSON MILLER

1

DESERT BLUES: HISTORICAL RIFFS

ON THE MOROCCAN PERIPHERY

THE DESERT AND HISTORICAL DISCOURSE

Recently I accompanied a group of Harvard alumni as their "study leader" on a whirl-wind tour of Morocco's imperial cities. My assignment was to add depth to their travel experience by interjecting nuggets of "expert" information at just the right moment. When we ventured out from our five-star cocoons and visited the wonderful sites that make Morocco's cities such architectural gems—the Tour Hassan minaret of Rabat, the Kutubiya mosque minaret in Marrakesh—I found opportunities to slip in some hard-core history. I tried to create in short sound-bites some of the historical context for appreciating these sites. I told them about the morphology and life cycles of cities, about the processes of urban planning, and about the great personalities—sultans, kings, proconsuls—that ruled over them. My stories seemed to hold their attention, generate new questions, and broaden their understanding. Soon I realized that I was inadvertently engaging in practices associated with the "spatial turn"—that is, taking the landscape beyond the merely pictorial and contextualizing it, giving history equal time with an appreciation of "the view."[1] Their response was enthusiastic. And so I came to see at first hand the power of narrative history in shaping the tourist experience.

How does one translate this particular insight into the desert setting? In the great imperial cities of Morocco, and for that matter, in any important city, we can readily identify places in the built environment that seem to evoke the past. But what visual stimuli in the desert do the same? True, there are the *ksars* and *kasbahs* that are remind-ers of a way of life that no longer exists. But these monuments, by the nature of their

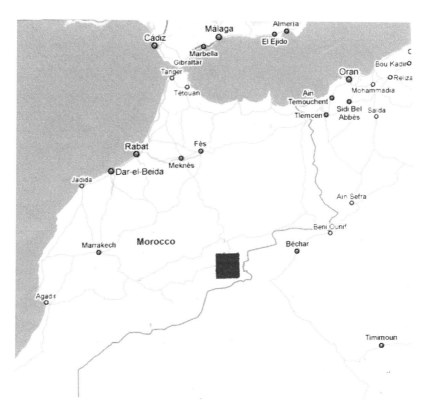

Map of Morocco showing the location of the Tafilalt region (© 2011 Google, map data ©2011 Europa Technologies, Google, Tele Atlas)

construction and their upkeep (or lack thereof), resist the effort to anchor them in time. In fact, they are decidedly ahistorical. One *ksar* looks very much like another, most are recently rebuilt, and those that are not are often depressing piles of rubble. Moreover, the manner of their construction reinforces sentiments that actually work against the grain of history, with their suggestion of timelessness and anonymity.

Our problem is to invent a historical discourse about the desert with properties of verisimilitude yet easily grasped by the nonspecialist. As the setting for this effort, I have chosen the Tafilalt region of southeastern Morocco, which has played a central role in Moroccan history. An important crossroads of trade and a gateway to the Sahara, it is also the homeland of Morocco's reigning dynasty, the Alawi family. The history of this region has been documented in both Arabic and Western languages, offering an array of source material containing themes of enduring social, political, and cultural relevance that, if read carefully, deepen our understanding of the desert way of life. In pursuing this brief analysis, I hope to make the argument that some familiarity with the desert as a subject of inquiry will enhance the tourist experience, in the belief that a historically and culturally informed visitor is a more valuable asset than an uninformed one.

THE CHRONICLES

The desert is often represented in histories of the royal family, the grand narratives that define the nature, quality, and success of the ruling dynasty. The tradition of dynastic historiography is very strong in Morocco, and each era has its chronicler(s) intent on adding to the glory of the leading family. The 'Alawi *sharifs*, or descendents of the Prophet, whose place of origin was the Tafilalt region, have ruled Morocco for the past 400 years. An ongoing theme in the dynastic chronicles written by court historians from the seventeenth to twentieth centuries has been the connection between the Tafilalt and the monarchy. They depict this desert oasis as the wellspring of the dynasty's power and the source of its primal strength, or *asabiya*. Although only three reigning Alawi sultans actually ever visited this remote region before the twentieth century, the spiritual tie between the desert and the imperial center was indelibly etched in the official historical mind and from there transferred to the learned classes in society.

There was yet another reason why the Tafilalt played a special role in the educated imaginary. It was also seen as a vital outpost and a key component in the construction of the idea of the national territory. The Tafilalt is depicted in the chronicles as defining the far reaches of the empire, while being very much part of it.[2] Facts on the ground supported this connection between center and periphery. For one thing, the tomb of the progenitor of the Alawi dynasty, Mawlay Ali al-Sharif, is located in the Tafilalt; it remains an important pilgrimage site. Moreover, the area near the tomb was populated with other members of the Alawi clan, both living and dead. Out-of-favor sons, fractious neph-

A *ksar* complex near Rich in the Tafilalt region

ews, and rebellious uncles were banished to the oasis to live out their days on subsidies from the palace, in a doubled-edged display of ruthlessness and generosity.[3] Thus the chronicles endow the Tafilalt with a multisided personality: holy burial ground, a place of refuge, frontier outpost, and gilded cage. For the royal historians, the region refracted qualities of piety, asceticism, and uncompromising authority that the monarchy was meant to represent, against the background of that pure primitive medium, the soil of the desert.

A particularly dramatic demonstration of this connection between palace and desert—and one that has contemporary resonance—took place at the end of the nineteenth century. The French at the time were exerting strong pressure on Morocco from their colonial outposts in southern Algeria. Surrounded by his entire court, his harem, and an army of 18,000 men, Sultan Hassan I marched to the Tafilalt from Fez—a distance of more than 600 kilometers—and remained there for several weeks. A visit to the tomb of Mawlay Ali al-Sharif was the centerpiece of his stay; it was staged as a colorful pageant, with horses flashing trappings of green and gold, thousands of mounted men, and green banners waving in the desert breeze. It must have been an unforgettable sight—a real tourist moment. It was all part of a *grande geste*, a political maneuver calculated to mark the limits of the state and to serve notice on the French that they were infringing on the Moroccan borderland.[4]

Oddly enough, this historical event has contemporary resonance. The notion of possession by preemption has been an ongoing theme in Moroccan history. We may recall the Green March of November 1975, when approximately 350,000 Moroccans converged on the city of Tarfaya on the edge of the Spanish Sahara and waited there for a signal from King Hassan II to cross. They brandished flags, photographs of the king, and copies of the Koran, and once again, the dominant color was green, the color of Islam.[5]

EXPLORERS IN DISGUISE

A second kind of riff on the desert are accounts of exploration and discovery by Europeans. These accounts are few and far between before the twentieth century, with some notable exceptions, among them the sixteenth-century account of Leo Africanus, or al-Hassan al-Wazzan, a Muslim Granadan who wrote the first geography of sub-Saharan Africa published in the West, entitled *A History of Africa*. Before he was captured by pirates in the Mediterranean and presented as a prize to Pope Leo X in Rome, Al-Wazzan criss-crossed the Sahara several times, recording data about human geography, trade routes, plants, and fauna. During a journey to Timbuctu in 1512, he passed through the oasis of Tafilalt, which he calls Sijilmassa. Al-Wazzan discusses many topics that were completely new to his European audience: the feuding among the Saharan tribes; the complex management of its water systems; the trade in gold and slaves. Al-Wazzan was a rare intellect for his time—a border crosser who lived in two worlds at once. He brought Africa into the consciousness of the Renaissance reader, broadening Europe's vision of the known world, and eventually he returned to North Africans precious information about their own societies not found in other sources.[6]

CHARLES DE FOUCAULD.

French explorer Charles de Foucauld (1858–1916). Dressed in the disguise of a Jewish rabbi, he traveled through the Moroccan south in 1883.

In the nineteenth century, more explorers went to southern Morocco and reported back to Europe. One of these travelers was Charles de Foucauld, a young French aristocrat who traveled across the Moroccan desert in 1883–84 in the guise of a Jew. His book, *Reconnaissance du Maroc*, published in 1888, is a rich compendium of geographical and ethnographic knowledge about all of Morocco, but especially about its desert zone.[7]

Because of the grace and elegance of its style, this source is a premier literary work, exposing the complex identities of the author. The disguise de Foucauld chose was ironic, for in his text he reveals no love for the Israelite nation. His own biases apart, his account was acclaimed for its erudition and accuracy, and immediately found a place in the expanding canon of literature about geographical exploration that was nurturing new epistemologies of colonial science.[8] The *Reconnaissance* is remarkable in the subtlety of its viewpoint, the mass of data it contains, and most of all, for de Foucauld's discovery of a desert society that was as complex, differentiated, and conflict ridden as his own. While he never actually reached the Tafilalt—the closest he came was Ferkla in the Todra river valley—his observations and concise notes became the gold standard of desert exploration, often emulated but rarely matched by later voyagers.

He was closely followed by Walter Harris, correspondent of the *London Times*, who spent much of his adult life in Morocco reporting on the political intrigues that led to the French Protectorate. Harris was another well-heeled aristocrat fleeing the boredom

French colonial railway poster from 1933 entitled "Gateway to the Tafilalt"

of Victorian London; in Morocco, he adopted the flamboyant style of the explorer and intrepid traveler. In 1893, he voyaged through the south, also in disguise, choosing as his persona that of the servant of a *sharif*. Wearing a tattered *jellaba*, a necklace of thick wooden prayer beads, and no shoes, he was accompanied by a Berber guard for protection. He managed to bamboozle the naïve Berbers into thinking he was a Syrian savant because of his fluid Arabic. On his return home, Harris wrote his reminiscences in a book entitled *Tafilet*, illustrated by his own drawings and photographs that were the first images of the oasis to be published in the West.[9]

Both Harris and de Foucauld were richly imbued with the prejudices of their age; Harris was fascinated by matters of race, with who was Berber and who was Arab, imagining the Berbers as "moral, simple and peaceful"—a much higher life form than Arabs, and closer, of course, to Europeans.[10] De Foucauld, despite his keen scientific understanding, believed the same. Both men saw women as mere beasts of burden and blacks as smiling and cheerful in their servitude. Arabs were "plundering thieves" and Jews were "despised" rapscallions, always looking for the best deal, although admittedly skilled at certain tasks, such as gunsmithy.[11] The amalgam of witlessness and wisdom displayed by explorers like de Foucauld and Harris, the stereotypes they

Entryway to a *ksar*
Door from the kasbah of Tiflit

inspired with their reports from the field, fired the European imagination and helped to shape the desert imagery slowly becoming canonized by the West. The influence of their ideas was decisive on the generation that followed, the colonizers who opened the Tafilalt to popular tourism.

DESERT BLUES

A third and final riff on desert historiography is what I call a local one, generated by post-independence native Moroccan scholars. This brand of historical writing presents a wholly new perspective on the desert and its place in the Moroccan experience. The best example of the genre is a book entitled *Le Tafilalt* by Larbi Mezzine, a thesis published by the Faculty of Letters in Rabat in 1987.[12]

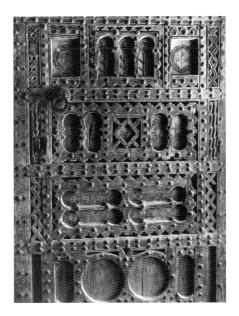

Drawings by Maurice Romberg, from sketches and photos by Walter B. Harris

"In Tafilet"

"The Kasbah of Sekoura"

"A Corner of a Sôk—Early Morning"

"Young Women of Tafilet Drying Dates"

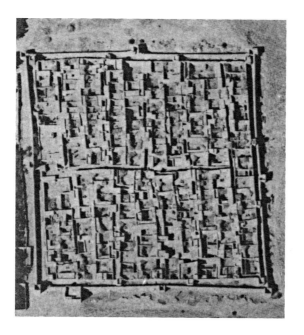

Aerial view of the *ksar* of Lgara, whose complex social life was revealed in a recently discovered notebook

As much as dynastic chronicles have a well-regarded niche in Moroccan historiography, so does the writing of local history. In fact, some of the most highly respected works of twentieth-century Moroccan historiography deal with specific regions of the country, such as the *Ma'sul* of Mukhtar al-Susi, and Mohammed Da'ud's *Tarikh Titwan*.[13] Although these older works are rich in documentation, they are also problematic, because they focus on the concerns of another age: elite power, changes in authority, the rise and fall of local governors, the virtues of local saints, the importance and prestige of their region as opposed to others.

Mezzine's work builds on this tradition but goes beyond it, adding new elements inspired by a consciousness of the ecology, the people, and the specific social reality of the desert environment. He has a sense of the strenuousness of life in the desert and the vulnerability of the social systems that sustain it. Moreover, he is reacting against the colonial historians who created some of the most tenacious myths about the Moroccan past, as well as against the postcolonial ideologues who saw history through the singular prism of the nation-state.[14] Instead, he concentrates on the concrete and the definitive, such as the daily problems of living in the parsimony of an oasis, the constant menace of marauding nomads, the stresses arising from the cohabitation of various groups in limited space.

One example will suffice to make this point: Mezzine came in possession of the ordinances of the *ksar* of Lgara in the Tafilalt dating from the second half of the nineteenth century, given to him by some high-school students in Erfoud.[15] They were contained in a small notebook of twenty pages: among the topics covered were the rules and regulations for life inside the *ksar*, the penalties for infractions, rules for the distribution of water, and rules governing the condition of the *Haratin*, the black field workers essential to the oasis economy. Written in a basic Arabic, the document, in Mezzine's words, "exudes the odor of the land in its minute details" and reveals the overwhelming human effort required for living in this closed environment.[16] One cannot read it without appreciating the value of the place about which it speaks.

This brilliant thesis, entirely original in its conception, demonstrates the point that carefully mapped historical trajectories, whether they focus on power or nature, people or places, provide stories that stimulate the historical imagination, enrich our understanding of place, and provide an element of truth to an otherwise disembodied and mystified experience. Blues and riffs, space and time interweave to form the dense harmonies of the lived moment. If the destiny of the Moroccan desert is to be commodified and rendered up as a tourist product—and no doubt it will be—then at least let it be done with intelligence and respect for the past and the people who once walked there, whether they were Berber oasis-dwellers, European explorers, or ruling sultans.

NOTES

1   The subject of the complex relationship between spatial and temporal narratives, a central concern of contemporary cultural theory, is beyond our scope. For a useful overview, see Edward Soja, "History, Geography, Modernity," in Simon During, ed., *The Cultural Studies Reader* (London and New York: Routledge, 1999), ch. 8. Historicist discourse in a tourist context is discussed in Dean MacCannell, *The Tourist: A New Theory of the Leisure Class* (Berkeley: University of California Press, 1999).

2   National territory can be defined by describing the space affected by a natural calamity: "Locusts appeared in the region of Marrakesh, and extended all the way to Salé and Tafilalt," wrote the eighteenth-century Moroccan historian Muhammad al-Qadiri. Norman Cigar, ed., *Muhammad al-Qadiri's Nashr al-mathani: The Chronicles* (Oxford: Oxford University Press, 1981), 130.

3   Ibid., 116, 129, 138.

4   Mawlay Hassan's visit to the Tafilalt in 1893 was documented most completely by Docteur F. Linarès, "Voyage au Tafilalt avec S.M. le Sultan Moulay Hassan en 1893," *Bulletin de l'Institut d'Hygiène du Maroc*, (July–Sept. 1932), 93–166; (Oct.–Dec. 1932), 95-136; Mohammed Aafif, "Les harkas hassaniennes d'apres l'oeuvre d'a. Ibn Zidane," *Hespéris-Tamuda* 19 (1980–81): 153–168. See also Walter Harris, *Tafilet: The Narrative of a Journey of Exploration in the Atlas Mountains and the Oases of the North-west Sahara* (Edinburgh: W. Blackwood and Sons, 1895), 255-257.

5   Pierre Vermeren, *Histoire du Maroc depuis l'indépendance* (Paris: La Découverte, 2002), 73–74.

6   Leo Africanus, *The History and Description of Africa* (New York: Burt Franklin, 1967), 782–785. For a recent study, see Natalie Zemon Davis, *Trickster Travels: A Sixteenth-Century Muslim between Worlds* (New York: Hill and Wang, 2006), Introduction. On Sijilmassa, see Dale Lightfoot and James Miller, "Sijilmassa: The Rise and Fall of a Walled Oasis in Medieval Morocco," *Annals of the Association of American Geographers*, 86,1 (March 1996), 78–101.

7   Charles de Foucauld, *Reconnaissance du Maroc, 1883–84* (Paris: Challamel, 1888).

8   Daniel Nordman, "La Reconnaisance du Maroc, de Charles de Foucauld," in *Profils du Maghreb; Frontiéres, figures et territoires (XVIIIe–XXe siècle)* (Rabat: Université Mohammed V, Faculté des Lettres, 1996), 141–180.

9   Harris, *Tafilet*.

10 Walter B. Harris, "A Journey to Tafilet," *The Geographical Journal*, 5, no. 4 (April 1895), 324, 326.

11 Ibid., 326.

12 Larbi Mezzine, *Le Tafilalt: contribution à l'histoire du Maroc aux XVIIe et XVIIIe siècles* (Rabat: Faculté des lettres et des sciences humaines, 1987).

13 Muhammad al-Mukhtar al-Susi, *al-Ma'sul*, 20 vols. (Casablanca: Matba'a al-Najah, 1960–63); Muḥammad Dāwūd, Tārīkh Tiṭwān, 10 vols. (Rabat: Kullīyat al-Ādāb wa-al-'Ulūm al-Insānīyah, Jāmiʻat Muḥammad al-Khāmis, 1959–1990).

14 Mezzine, *Tafilalt*, 16.

15 Ibid., 34–44, 335–360.

16 Ibid., 36.

CLAUDE PRELORENZO

2

THE DESERT AS A (MOVIE) MYTH

Few of us have ever seen the desert, yet we all have in our minds a clear and fascinating image of it. Is it not the very characteristic of a myth to convey a reality, a phenomenon, an appearance—or a "form," as semiologists would say—that is shrouded in mystery? Much of our mental representation being cast as images, the myth is reinforced not only through literary descriptions but especially by pictorial and photographic media, and since the beginning of the twentieth century, by movies. Because of their meaning and usage, some myths (related to fertility, the mother, the cosmos, etc.) are universal. As cultural phenomena, however, myths are rooted in historical and sociological contexts. This is why I would warn readers that most of the analysis to follow is based specifically on the Western myth of the desert. It is important to clarify this limitation, as this myth arose at the same time as the desert was being tamed via colonial dominance and commercial exploitation, and is therefore closely related to the idea of conquest. In the present fluctuating balance of world power, we need to know to what extent the new status of desert territory (and society) will change this cinematographical myth, even if cinema remains almost entirely under Western influence.

Most of this discussion will be devoted to hot, dry deserts (I will not attempt to see whether polar deserts obey the same rules), and the Sahara will be the focus, not only because it is so present in our minds and has been visited and written about so much, but also because it is the most mythical desert. Placing the Sahara at the center of concern also implies taking into account the Muslim and Arab civilization that is part of its context. It is likely that the North American deserts, linked to American Indian contexts, African ones inhabited by animist cultures, or Asian deserts associated with nomads and Eastern empires constitute a broader group of diverse and familiar mythological families that would all merit being studied. However, that is not my purpose here.

THE CONSTRUCTION OF THE MYTH

I will base my analysis of the cinematographical myth of the desert on two theorical propositions: that of the Romanian philosopher of religion Mircea Eliade, and that of the French semiologist Roland Barthes. In his major work, *Cosmos and History: The Myth of the Eternal Return,* Eliade considers that the reality of an object or an event can only exist inasmuch as it results from the repetition of an action dating back to a primordial mythical era.[1] The whole meaning is encapsulated in this primary myth. Contemporary anthropologists have adopted the idea that the "genetic" function of the myth is that of a primal language. In this context we have used two criteria: origin and repetition, in which the literary form is considered the original one, and the cinematographical expression a reiteration.

In *Mythologies,* Roland Barthes proposes the idea that myths can be analyzed from the semiological point of view, but not linguistically.[2] Inasmuch as myths are part of a discourse in the broad sense of the term, above all they are "forms." Hence the sense of the myth (the "form") corresponds to the social significance of the field or the object whose meaning it has taken.

LITERARY CORPUS

A few well-known literary works have contributed to create the myth of the desert:

> *Le désert* (1895), Pierre Loti
> *The Seven Pillars of Wisdom* (1926), T.E. Lawrence
> *Il deserto dei Tartari* (1940), Dino Buzzati
> *Le Petit Prince* (1943), Antoine de Saint-Exupéry
> *The Sheltering Sky* (1949), Paul Bowles

Tourist guides such as the famous Baedeker and the Guide Bleu are also useful additions to our literary traveling bibliography.

Pierre Loti's *Le désert* relates the crossing of the Sinai by caravan from Suez to Gaza between February 22 and March 25, 1894.[3] This work is highly significant in France and Europe, not only because of Loti's reputation but also because it advances most of the primary components of the myth. Loti's writing can be considered a kind of thesaurus of meanings of "desert." As a founder of desert literature, his primeval "images" will later be used again and again in written works and films. This crossing of the desert—except for a stay in the monasterial fortress of St. Catherine (the "burning bush" basilica) in the heart of Mount Sinai, as well as a few stops in oases—is first and foremost a long contemplation of the landscape, the mineral universe, color and light (Loti was also a painter) that mixes an aesthetic and somewhat precious vision of the desert with a more tense evocation in which man is confronted with nature in its least cultivated form.

The themes retained by Loti, in the form of a diary, cover a broad spectrum:

1. The desert is above all a spectacle whose beauty, light, colors, and clean air beguile him. "It is like the apparition of a magical world" (p. 78). "The enchanting paradisial light radiates everywhere" (p. 119). "Evening has come, the golden hour....gold is spread out limitlessly" (p. 111).

2. The desert exists through its emptiness, its monotonony, dominated by infinity, without any horizon and the immensity of its space. "The impression of the desert" derives from "the perceivable assertion of its immensity" (p. 35). It is "colourless, sorrowful desolation" (p. 78). "Confronted with such emptiness, an anxiety rises within us corresponding to our own." Contrary to the "men in tents," "the man from the house of stone that lies buried in the depths of our collective consciousness due to our atavistic tendencies is vaguely anxious to find himself without a roof, no walls and to realize that none can be found in the vicinity" (p. 46). The desert is compared to "the emptiness of the sky" (p. 78) and of the sea, "monotonous as the sea" (p. 79). "The desert of Tih's loneliness (....) is as immense and flat as the sea" (p. 145). "The plain, reduced to its essence, is little more than a space in which our eyes encounter nothing" (p. 171), "and it is always silent and always lifeless" (p. 46).

3. The desert is the place of the dead, a finite world. "It is as though we are in a finite world scorched by fire that no dew will ever fertilise" (p. 83). "All is emptiness, silence and death" (p. 36). "Such is the splendour of these unvarying regions" (p. 36). "Islamic immobility and the peace of death are spread everywhere" (p. 115). "The Orient made eternal in its dreams and its dust" (p. 145), "this Arabia that has neither earth nor plants, only bones" (p. 90).

4. The desert is the site of genesis and eternity. The desert represents the permanence of eternal matter. "One would think we are witnessing some grandiose silent spectacle from the eons of geology" (p. 105). "Such is the splendour of these unvarying regions from which are absent these ephemeral delusions, forests, vegetation and grass; it is the splendour of quasi eternal matter liberated from all the instability of life ; geological splendour before the creations" (p. 36). "Such places, as yet undefiled by man or nature, are necessary to allow us to conceive, we, so small and taken with minutiae, what the creation of the world may have been like at its horrifying grandiose birth" (p. 95).

5. The desert is preeminently a virginal place. "It is a cursed part of the earth, that wishes to remain unpenetrated, where man is not welcome" (p. 80). "Mountain chains are interlinked and heaped on top of each other with regular shapes that, since the beginning of the world, have remained untouched by all human intervention" (p. 35). Confronted with this virginity, man transforms and cleanses himself, as at the beginning of time "to emerge from one's tent in the splendour of a virginal morning; relax ones arms, stretch out half naked in the pure cold air; wind one's turban and drape oneself with a white veil of wool...simply to be alive" (p. 34). But, "after the desert, the Promised Land" (p. 185).

The second inspirational text, which unlike Loti's will inspire many films, is that of the story of *Le Petit Prince* by Antoine de Saint-Exupéry.[4] Whereas Loti's writing is full of descriptions, Saint-Exupéry's is more of a parable than a description. He nevertheless succeeds in creating a powerful spiritual vision of the desert. The desert in which the pilot (the author himself) breaks down is a place where he makes an important discovery: his reality and scale of values are not the conventional ones of adults but the poetically naive ones of children. The little prince will help the pilot reunite with the child in his soul and reconnect him with the true values of love. This revelatory power of the desert is ever-present in the Judeo-Christian tradition (contrary to other religions founded in caves, forests, or mountains). The crossing of the desert as an ascetic trial, a purifying experience, in the Old Testament: Christ's retreat into the desert and the reclusive lives of hermits; the "desert" as an expression of clandestinity for the persecuted.

Some lesser notions can be found in Saint-Exupéry in addition to the desert as a revelatory place. These reinforce ideas already expressed by Loti:

- The distance from inhabited regions ("I found myself a thousand miles from any inhabitants" (p. 16).
- The solitude ("Here is the desert. There is nobody in the desert, said the snake," "Where are all the human beings? asked the little prince. It is a bit lonely in the desert" (p. 64).
- The loss of direction, the idea of having "gone astray" (p. 16).
- The desert as a threat to life itself (fear of "dying of weariness...of hunger...of thirst...of fear" (p. 16) refers also to the drawing (p. 61): dunes, sun, cactus, bones.
- The comparison with the sea ("I was more alone than a shipwrecked sailor on a raft in the middle of the ocean" (p. 15).
- The desert as an aesthetic spectacle ("I've always loved the desert. You sit on a sand dune. You see nothing. You hear nothing. Yet something is silently shining" "whether it is a house, the stars or the desert, what makes them beautiful is invisible" (p. 82) "at first light the sand is the colour of honey" (p. 85), "let's go and see a sunset" (p. 30).

The Seven Pillars of Wisdom by T. E. Lawrence, published in 1926, is not, strictly speaking, a book about the desert but more of an introverted, somewhat blunt, biographical essay.[5] Nevertheless, the desert plays an important role because it is the backdrop against which Lawrence realizes his military, social, cultural, and, it would seem, sexual destiny. This work, along with *The Little Prince,* constitutes the second example of spiritual desert literature.

Between the two world wars, desert literature evolved. The human butchery of World War I, the Great Depression, the rise of fascism, and the advent of World War II gave rise to doubts about humanity's very essence. Here began a human identity crisis in which new forms of nihilism emerge: the fear of inhumane absurdity as predicted by Kafka (*The Trial,* published in 1925) and relayed by Sartrian existentialism (*Nausea* in 1938, *Being and Nothingness* in 1943). Two other novels reflect this attitude: the first is

T.E. Lawrence in Arab dress, photographed by Harry Chase, 1918 (reprinted by permission of The Imperial War Museum, image Q73535)

Dino Buzzati's *The Tartar Steppe* (Il deserto dei Tartari), published in 1940, the second, the American Paul Bowles's *The Sheltering Sky,* published in 1949 and translated into French under the title *Tea in the Sahara* (the heading of the first of the three chapters of the book).[6]

In *Il deserto dei Tartari,* Buzzati analyzes the behavior of a group of soldiers posted in a fort on the northern frontier of a non-identified country, overlooking a wide, empty, sterile landscape in which nothing has ever happened. The timeless fort, deprived of any contact with the outside world, relies exclusively on repetitive military rituals, and the lack of initiative gradually drags the men relentlessly down into an uneventful existence until they become fixated on the advent of death.

Paul Bowles scans the desert to search for the meaning of existence. Although his fascination and aesthetic contemplation are evident, the experience of the desert and the quest of the desert are, for Port, the husband, a path toward death and for Kit, the wife, a path toward life. Port is a "traveler" as opposed to what Bowles calls a "tourist": "A tourist is somebody going to a certain place. The moment he has arrived, he just dreams to be home. The traveller is somebody who once he has got to the place he wants to go, decides to go further and further. A traveller is somebody that, ideally, will never be able to be back home."

In the end, Port will never return home, but neither will he reach his goal of going as far as possible into the desert, where he could get closer to reality. "The landscape was there and he thought more than ever that he could reach it. The rocks and the sky were everywhere, ready to absolve him, but he still had this inhibition within himself weighing on him" (p. 171). "I kept thinking that I would be able to penetrate it somehow. But when I think I am getting close, I get lost" (p. 170).

Kit watches over him rather than accompanying him. Not until he is dying from typhoid is she capable of touching him. From that moment, the desert that had been nothing but dust, heat, vermin, filth, disgusting food, and sordid hotels was revealed to her in all its splendor, serenity, and purity, with rose-colored dunes and blue waters in the oasis in which she bathes and is reborn. "Looking at the tranquil garden, she suddenly has the impression that she had never seen anything so clearly since her childhood. Suddenly life was there and she could plunge right into it instead of being content with looking at it from the window" (p. 252). In the end she is reborn bodily and sexually by becoming the consenting prisoner of a veiled Touareg.

Religious writings have also long evoked the desert as a particularly mystical place, as well as one of trials and rebirths. In the "crossing of the desert" in the Old Testament, Moses loses his princely status. (In the film version, Yul Brynner says to Charlton Heston: "The desert is your new kingdom.") The desert tests the faith of Israel : "This is where he tests them" (Exodus). In the New Testament, Jesus is quoted as saying "I am a voice screaming in the desert," tempted by the devil in the desert.

CINEMATOGRAPHICAL CORPUS

Numerous films are set in deserts. I could have made a long list if I had wished to see to what extent the myth of the desert had fed the film industry or to analyze their themes and identify their references. However, the objective here is not this kind of knowledge, what I call a "sociological" study; rather, the aim is to understand the process of creation of the myth through films, not how it has spread. To do so, only a few well-chosen examples from the history of cinema will be necessary to illustrate the main themes that have infiltrated and deepened the myth. These are:

The Sheik (Valentino), 1921, George Melford
The Son of the Sheik, 1926, George Fitzmaurice
Sahara, 1943, Zoltan Korda
Lawrence of Arabia, 1962, David Lean
The Little Prince, 1974, Stanley Donen
The Passenger, 1975, Michelangelo Antonioni
The Sheltering Sky, 1990, Bernardo Bertolucci

In a few films, the desert is used periodically as a backdrop for climactic scenes:

The Ten Commandments, 1956, Cecil B. DeMille
Zabriskie Point, 1970, Michelangelo Antonioni
Syriana, 2005, Stephen Gaghan
La Piste (The Trail), 2006, Eric Valli

In some other films, the desert is used more as a backdrop than as a central theme, but their pictorial strength has nevertheless contributed to constructing the Western myth of the desert:

Greed (Les rapaces), 1924, Erich von Stroheim
La Bandera, 1935, Julien Duvivier
Beau Geste, 1939, William A. Wellman
The Treasure of the Sierra Madre, 1948, John Huston
The Desert Rats, 1953, Robert Wise
Il deserto dei Tartari, 1976, Valerio Zurlini
Gerry, 2002, Gus Van Sant

Of course there are many others, especially Westerns, such as John Ford's films, and some science-fiction movies, such as *Dune* (1984, David Lynch) or *Star Wars* (1977, George Lucas).

COMPONENTS THAT MAKE UP THE CINEMATOGRAPHICAL MYTH

Films are bearers of ideas anchored in language, culture, and literature, and as such they propagate groups of mental images that create the myth of the desert in the imagination of mass culture. Four of these are particularly significant: the legend of the desert; the inhumanity of the desert; the desert from the outside as an expression of the inside; and desert peoples. These groups are complementary, but as society evolves they become differentiated. Cinema as a mass medium is predetermined by its economic nature, which is continually adjusting to the diverse sensitivities of its audience. At the beginning of the last century, it appears that what mattered most to the audience was the myth of the legendary desert, that of the Arab as in *The Sheik.* Between and immediately after the two world wars the sky darkened, and the desert became a symbol of a hostile environment in which man's integration could be only temporary, at the cost of a fight for survival that would shape his character, as in *Sahara.* In the more "psychological" 1960s and 1970s, films focused more on characters whose feelings and tensions are revealed by the desert, as in *Lawrence of Arabia, The Passenger* (original title *Professione: Reporter*). In recent years the desert, like the mountain, is presented as a resource for humanity that should be respected, protected, and "listened to."

In this evolution there are some invariables—a kind of backbone of the myth— onto which can be grafted somewhat more circumstantial attributes. The aesthetics of the desert are always resplendant with purity, beauty, and a certain grace, as described in detail by Pierre Loti and as can be found whatever the context, even if incompatible with human life. Another common factor is spirituality. In the films mentioned, the desert serves the intrigue of the story by transcending it. The solitude, the virginal aspect of the place, and a certain formal unadulterated perfection as compared to civilized territory leads the characters to an elevation of their thoughts and behavior.

THE ARABIAN DESERT LEGEND

The legend of the desert is the most ancient and most common aspect of the myth—it is the foundation of the modern image of the desert. This legend is mirrored without exception in the background of all these films, even those of Antonioni, Bertolucci, and Van Sant. In the worst cases it is a stereotype, in others an archetype, but in all cases it creates a cultural complicity with the spectator. The legend of the desert in the Western world goes hand in hand with the phantasmagorical Arabia of the nineteenth and early twentieth centuries, when a certain distance facilitated the free expression of oneiric fantasies or pure fabulation. The roots of this legend can be found in the fascination for the Orient shown by Western artists such as Ingres, famous for his *Turkish Bath* and Delacroix's *The Death of Sardanapalus* (inspired by a poem by Byron about an Oriental prince steeped in luxury, voluptuousness, and cruelty). From a literary angle, Pierre Loti is the best example of an author infatuated with the Orient. On a somewhat higher literary scale, *Salammbô,* by Gustave Flaubert, and *Le roman de la momie* by Théophile Gauthier, should be noted. At the time of French colonial expansion, toward the end of the nineteenth century, Arabization was dominant. It was propagated by individuals such as Général Lyautey, the French military governor of Morocco, who set out to protect Islamic culture and sought a path for modern architecture inspired by Arabo-Islamism.[7] Just a few years before his death in 1934, he accomplished his mission by organizing the colonial exhibition of 1931. Edward Said found fault with Orientalism, which he criticized as being "nothing better than a Western way to dominate, reorganize and impose itself on the Orient."[8] According to Conrad Phillip Kottak, Westerners "will attempt to remake the native culture within their own image, ignoring the fact that the models of culture that they have created are inappropriate for settings outside Western civilization."[9]

Another facet of this legend is that of the Arabian city, magical and childlike, as depicted in *The Thief of Bagdad: An Arabian Fantasy in Technicolor,* along with films inspired by *A Thousand and One Nights*, such as *Aladdin and the Magic Lamp, A Thousand and One Nights,* and *Sinbad the Sailor.*[10]

The legend is also composed with human figures, always in an idealistic and "picturesque" mode. Paintings, books, and films picture an imaginary desert featuring angry lions, feverish horses, women lingering in steamy hammams or under some sort of attack, where walls collapse under the weight of tapestries and water drips off mosaics, where intrigue is dramatic and violent, and the image of the Queen of Sheba (Gina Lollobrigida) is mixed up with that of Cleopatra (Elizabeth Taylor) and Salome (Rita Hayworth).[11]

THE HOSTILE DESERT

However legendary, the desert is nevertheless represented on film as a particularly foreign environment—hostile for humanity, without water, plants, or animals except for some poisonous beasts, lacking shade, burning in the day and glacial at night, prone to violent sandstorms, with each of these characteristics making its tragic contribution to the scenario. Moses (*The Ten Commandments*) bent over by violent sandstorms, suffers from thirst; Humphrey Bogart (*Sahara*) accomplishes a miracle in an epic scene by crossing the Libyan desert without a drop of water; Jack Nicholson (*The Passenger/ Professione: Reporter*) struggles to reach the nearest village from his broken-down vehicle, while the airplane pilot (*The Little Prince*) wastes his precious time and supplies on a strange child; and Peter O'Toole (*Lawrence of Arabia*), to the consternation of his entourage, sets off in search of the lost Gassim, above whom vultures are swirling in a contrived image evoking the imminence of death.

This threat of the hostile desert is an integral and important part of the myth. It serves as a dramatic backdrop and is emphasized in every film. However, in more recent examples, such as *La Piste (The Trail)*, this hostility is somewhat tempered as the desert can be tamed; the little girl and her guide are resigned to death in the vast, sterile sand dunes when the desert reveals its hidden treasure. The guide knows his territory and finds, buried in the earth, invisible to an uninitiated eye, a plant that contains a life-saving liquid. In the present context of "ecological" cinema, the theme of the desert, like that of mountains, polar regions, and swamps, has a growing positive connotation. It is being portrayed as an ascetic place in communion with nature, full of hidden riches. In these depictions, the desert can no longer be vanquished as it was in *Sahara* or *Lawrence of Arabia*, but truth can be revealed by surpassing one's limits, by confronting death.

THE INTERIOR DESERT

Like all representations, the myth of the desert in cinema is not only a place but an allegory, a metaphor for an interior space or moral landscape. The question arises: what contribution does the choice of dramatic place make to a film? The first answer to this question is linked to the romantic nature of a film. Places are used to reflect feelings—the expression of a frame of mind. This is similar to the syndrome of *Le Lac*.[12] At night Lawrence retires to the desert to meditate and calm his concern about not knowing what to do. In the morning, the sun rises behind the sand dunes and as life returns, the solution becomes clear: the town of Aqaba has to be taken from the Turks, from the side facing the desert, the defenseless side, which the invaders thought was invincible. This same sequence illustrates another function: the retreat in the desert. The idea of the retreat—that is, retiring from the world of humanity—is common in religious and ascetic practices. The retreat in the desert is especially prominent in Jewish and Christian traditions. It privileges a desert synonymous with purity and complete emptiness, which facilitates communicating with the heavens. The desert is a solitary place, in which the person, because he is alone, is face-to-face with his beliefs and discovers his destiny.

Another way the cinema responded is brilliantly related by Céline Scémama-Heard in her book about Antonioni.[13] Evoking *The Cry* (*Il Grido*, Michelangelo Antonioni, 1957), the author quotes Richard Roud, who considered that the "intrigue of the film—including the feelings of Aldo—could be said to be the expression of the landscape rather than the reverse." Scémama-Heard continues: "It is not a question of rejecting these concepts nor of preferring one to the other. Wishing to 'enclose' the spatial role in a function, be it analogically to the role or independently of the role… inhibits us from identifying what these categories represent." She remarks that in *The Passenger,* "just as the car gets stuck in the sand, the camera abandons it and with a long slow panoramic shot reveals the desert in all its immensity. Obviously the aim is not to be descriptive; its function is not to show us the person in the landscape, in fact the character is out of sight as though he is set apart and therefore does not belong to the place where he is standing. The result is that Locke seems to be cut off from the real world."

In its relationship with people, the desert, as portrayed in movies, plays many different roles: a situation (familiarity/externality), an itinerary (the trial of going into the unknown), a revelation (ascetic), and a relationship to the world.

DESERT CHARACTERS

The myth of the desert is not only made up of the spectacle of nature and the expression of feelings; it is also forged from the people who live in it, use it, witness it, and express it. The first and most important character to symbolize the legend of the desert is that of the Sheik, the Prince of the Desert, as played by Rudolph Valentino in the 1921 film of the same name by George Melford. The character will reappear in his last film in 1926 under the direction of George Fitzmaurice.

These performances defined an icon that would long survive: a tall, handsome, brave, strong nobleman, more physical than intellectual, both generous and cruel and therefore alluring yet sexually troubling. This portrayal of non-European people is typical of a racist mode of thought. The myth without the Prince of the Desert would lose a lot of its power, so we have to look into the cinematographic history behind this recurring figure.

THE AL JOLSON FACTOR [14]

Yet another Westerner, forty years after Valentino, would take on the role of the Prince of the Desert: Peter O'Toole in *Lawrence of Arabia*. In Lawrence's case there is no ambiguity about his ethnic origins, but his disguise is identical to Valentino's. The distribution of parts is still done according to racist Hollywood rules; the only Arab actor to be cast among the major roles, in this case Egyptian Omar Sharif, gets low billing as Sherif Ali, compared to the other "Arabs" as played by Anthony Quinn, Alec Guinness, and José Ferrer. [15]

From a cinematographical point of view, this phenomenon is hard to explain. Probably it reflects a certain level of ignorance among the producers and directors of the Arab cinema world, and of its actors. But it is also based on realistic marketing—the fear that the public might reject characters with whom it does not wish to identify. Even stranger are the Touareg scenes from *The Sheltering Sky* of Bernardo Bertolucci. In the last sequence of the film, Kit tries to find a place in the real world by seducing a Touareg whom she views as solidly and legitimately planted in reality, unlike her equivocal husband or the travelers from her own country. For the role of the Touareg, Bertolucci chose a French-Vietnamese dancer from the Paris Opera House, Eric Vu-An. This is even more surprising in the context of Italian cinema, in which realism has always played a central role. It is true that a film creates its own reality, and cinematographical representation deploys all kinds of artifices that compose the vision presented to viewers. In a less important film, *Khartoum* by Basil Dearden of 1966, the role of Mahdi is played by a ridiculous eye-rolling, smirking Laurence Olivier. This cinematographical disguise has a parallel in the camouflage aesthetic that can be found in Pierre Loti's house, built in Rochefort, his birthplace. In a banal building, characteristic of this military town, he created an "Arab" interior where he had himself photographed in Bedouin costumes brought back from his travels. These well-known images are the first in the Loti/Valentino/O'Toole tryptic.

THE MYTH TODAY

Is the myth of the desert that was engendered through nineteenth-century literature and twentieth-century cinema the same today? The question has become complex as a variety of images are interwoven into present-day representations, including television documentaries. It would be interesting to look into the National Geographic channel, in particular, which has transferred to television the aesthetic portrayal of the desert previously presented in its journal. Further information and the ongoing propagation of the myth of the desert is provided by the internet. The myth as a social and intellectual phenomenon takes time to elaborate and is durable because a large number of its components are induced by unconscious mechanisms.

This can be seen in the film *La Piste (The Trail),* in which ecological factors coexist with more traditional signifiers: the aesthetization of the desert, asceticism, ethnic configurations, open spaces, etc. New parameters have been grafted onto the old.

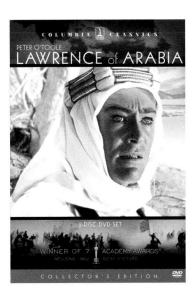

One is the disappearance of other types of virgin territory in favor almost exclusively of the desert. Over recent centuries, every country, region, and town possessed places reputed to be "wild," incompatible with human civilization and appropriated through myths in fairy tales, legends, and other kinds of storytelling: forests occupied by wolves or pumas, impenetrable woodlands inhabited by trolls and other weird creatures, stories of mysterious disappearances, wizards, buried treasure, fabulous mines, werewolves and vampires, etc. Gradually these places have disappeared as forests have been cleared, land colonized, and boundaries established. Europe has practically no trace left of primal forest. Only the largest countries such as Russia, the United States, and Brazil still have large tracts of natural land within their frontiers. The force of attraction for "real" natural spaces such as mountaintops, polar regions, the ocean depths, the Amazon forest, and of course, the desert, has been strengthened. The demand for a mythological world has therefore increased as the supply has decreased. These places are more distant in miles, though it takes less time to reach them, so they have become both more protected and more visited.

Another criterion derives from changes that have objectively and subjectively affected deserts in the twentieth century. Objectively, the desert is easily accessible, almost everywhere, by SUV's, helicopters, and planes; it is a source of subterranean riches: minerals, oil, precious stones. New capitals of the world are rising out of the desert as major business hubs, places of learning and culture. At last, prehistoric sites and the art of desert peoples are being recognized. This inhospitable, hostile, sterile, backward, uncultivated land has become a comfortable, luxurious tourist resort, high-tech and fashionable, at once the cradle of humanity and the foundation for the future. These mutations are modifying the mental image of the desert. For the general public, influenced by international media, it would appear that the myth oscillates between two poles: the modern desert and the legendary one: *Syriana* and *Lawrence of Arabia*.

—Translated from the French by Helen Grant Ross

NOTES

1 Mircea Eliade, *Cosmos and History: The Myth of the Eternal Return* (New York: Harper Torchbooks, 1959).

2 Roland Barthes, *Mythologies* (Paris: Editions du Seuil, 1957).

3 Pierre Loti, *Le desert* (first edition 1895).

4 Antoine de Saint-Exupéry, *Le Petit Prince* (Paris: Editions Gallimard, 1999; first edition in French and first edition in English, New York: Reynal and Hitchcock, 1943).

5 T.E. Lawrence, *Seven Pillars of Wisdom: A Triumph*, first private edition 1922; first public edition 1926.

6 Dino Buzzati, *The Tartar Steppe* (Boston: David Godine, 2005; first published in Italian as *Il deserto dei Tartari*, 1940); Paul Bowles, *The Sheltering Sky* (New York: New Directions, 1949); French edition, *Un thé au Sahara* (Paris: Editions Gallimard, 1952).

7 For more information on this topic, see François Beguin, *Arabisances: Décor architectural et tracé urbain en Afrique du Nord, 1830–1950* (Paris: Dunod, 1983).

8 Edward Said, *L'Orientalisme: L'Orient créé par l'Occident* (Paris: Editions du Seuil, 2005; first edition in English, 1978).

9 Conrad Phillip Kottak, *Window on Humanity* (New York: McGraw-Hill, 2005).

10 *A Thousand and One Nights* (dir. Alfred E. Green, 1945, with Cornel Wilde in the leading role; *Sinbad the Sailor* (dir. Richard Wallace, 1947, with Douglas Fairbanks Jr., Maureen O'Hara, and Anthony Quinn).

11 *Solomon and Sheba* (dir. King Vidor, 1959); *Cleopatra* (dir. Joseph L. Mankiewicz, 1963, and dir. Cecil B. DeMille, 1934, with Claudette Colbert); *Salome* (dir. William Dieterle, 1953).

12 "Le Lac" is the title of one of Lamartine's poems that is probably the most emblematic of the French romantic movement. "Oh temps, suspends ton vol ..., Que tout ce qu'on entend, l'on voit ou l'on respire/tout dise 'ils ont aimé'."

13 Céline Scémama-Heard, *Antonioni: le désert figuré* (Paris: L'Harmattan, 1997).

14 In *The Jazz Singer* (Alan Crosland, 1927), Jolson blackened his face with boot polish.

15 It is curious to note that Steven S. Caton, in his very thorough work *Lawrence of Arabia, A Film's Anthropology* (Berkeley: University of California Press, 1999), tried to counter this criticism when he noted that "Omar Sharif was not the only non-Westerner" (p. 55), citing three others (Gamil Ratib, I.S. Johar—who is not Arab, but Indian—and Zia Mohyeddin), yet failing to indicate that none of these actors was given a major role.

NEIL LEVINE

3

A UTOPIAN SPACE BETWEEN TOURISM AND DEVELOPMENT: FRANK LLOYD WRIGHT'S ARCHITECTURE FOR THE DESERTS OF THE AMERICAN SOUTHWEST

Frank Lloyd Wright was unusual among modern architects in having maintained his office for more than two-thirds of his career outside of a metropolitan area. And of those nearly fifty years, more than twenty were spent in the Sonoran desert at the winter headquarters he began building in 1938. At that time, Taliesin West was located well beyond any development, on unincorporated land 17 miles east of Phoenix, which then had a population of fewer than 65,000 people. The even closer town of Scottsdale, of which Taliesin West later became a part, had fewer than 1,000.[1]

The desert profoundly affected the character of Wright's work as he sought to develop forms that could be seen to have grown out of the harsh, arid, mountainous landscape and to create a living environment integral to it and compatible with it. But even more important for the overall purposes of this volume are (1) the fact that Wright's earliest contacts with the deserts of the American Southwest in the 1920s were related to the nascent tourist industry; and (2) that his own later winter headquarters presented itself as a kind of utopian space somewhere between an elitist, controlled form of tourism and the indiscriminate, seemingly uncontrollable real estate development that inexorably followed it in much of the Southwest. Indeed, the sprawl of single-family houses nearing the southern edge of Taliesin West's property by the early 1990s has now curled around the Wright buildings up to their eastern, southern, and western flanks.

Taliesin West (Wright winter headquarters), Scottsdale, Arizona, Frank Lloyd Wright, begun 1938.
Aerial view, looking southwest, c. 1946–47.

Taliesin West. Aerial view looking southwest, 2005.

WRIGHT IN DEATH VALLEY

Wright was not the only artist or writer during the 1920s and 1930s to become fascinated by the desert, as the careers of Georgia O'Keeffe, D. H. Lawrence, Edward Weston, and Willa Cather bear out, but he was the only one of significance to engage with the tourist industry in a direct, interactive way. Tourism in the Southwest, in any meaningful sense of the word, dates from the first decade of the twentieth century, when the Atchison, Topeka, and Santa Fe Railway joined with the Fred Harvey Company to create hotels, restaurants, and Native American craft gift shops along its route. The focus was less on the desert landscape as such than on the Indian populations that still were its primary inhabitants. Railroad travel, however, was very restrictive and it was not until the later 1920s, with the arrival of the automobile, that large tourist groups could be taken into the desert to visit Native American pueblos and villages. The Harvey Company called its popular day trips "Indian Detours."[2]

But even beyond the range of the Harvey Company lay Death Valley in the Mojave Desert, the mythic desert of American legend, named by the Forty-Niners who were able to escape alive after taking a misguided detour through it during the California Gold Rush. Located on the border between California and Nevada at the lowest and hottest point in the United States, Death Valley was the site of Wright's first venture into the desert and of his first design for a desert building. The commission, from the wealthy Chicago insurance company magnate Albert M. Johnson, was for a winter retreat for himself, his wife, Bessie, and his friend, protégé, and local guide and informant, Death Valley Scotty, the colorful huckster who had made a life for himself in the Valley claiming to have discovered a gold mine and getting gullible businessmen like Johnson to grubstake him and eventually house him in splendor.[3]

When Wright was taken to visit the site in the early months of 1924, there was no road in Death Valley—that only came two years later. Wright's introduction to the Mojave was thus to a nearly totally untouristed desert environment, but one that would change radically over the next few years, so much so that it was declared a National Monument in 1933 and by then boasted two hotels plus a major tourist attraction, namely Scotty's Castle. Hollywood certainly had something to do with this change. The film *Death Valley Scotty's Mine* was released in 1912, followed three years later by *Life's Whirlpool*, based on Frank Norris's novel *McTeague* (1899), whose devastating final scenes of desolation and murder take place in Death Valley. But the movie that truly dramatized the Valley and made its image among the most powerful natural scenes ever recorded on film was Erich von Stroheim's remake of *McTeague* under the title *Greed*, released in 1924. To represent the anti-hero's descent into the depths of depravity in the most realistic way possible, von Stroheim brought his actors and crew to the desert site for six weeks in the middle of the summer of 1923 where, in temperatures reaching well above 150 degrees, the actors playing McTeague and Marcus became so nearly delirious that they fought one another as if actually to the death.[4]

Six months after the shooting of *Greed*, Wright was driven by Johnson in his automobile over sand and dirt tracks to the place on the northeastern edge of Death Valley

Johnson Ranch, Death Valley, California, F. W. Kropf, 1922–24, state of construction c. 1924.

Johnson Desert Compound project, Death Valley, California, Wright, 1924–25.
Preliminary perspective, from south.

Opposite: Johnson Ranch site and Scotty's Castle (Matt Roy Thompson, C. Alexander MacNeilledge, and Martin de Dubovay, 1926–31). View from above, looking across Death Valley to Panamint Mountains.

where Johnson and Scotty had built several temporary structures and were now planning to make something more permanent and grandiose of them.[5] Known as Grapevine Canyon, the elevated site was fed by underground springs and could thus support a certain amount of agriculture. Wright incorporated the existing buildings into his project, encasing them in continuous, horizontally delineated, low-lying walled precincts of reinforced concrete block and linking them to one another and to the landscape by terraces and causeways built of the same material. More like an earthwork than a series of buildings, the architecture adopted what Wright later described as the desert's characteristic "striated and stratified" masses of "nature masonry" to define its relationship to the site.[6] But most important for this relationship was the asymmetrical site planning based, for the first time in the architect's work, on the use of a non-orthogonal, 30-60 degree geometry. Again, Wright maintained that such a geometry was precisely what defined the angles of the slopes of the surrounding desert mountains as well as the shapes of the alluvial fans that formed their base.[7]

Furnace Creek Inn, Death Valley, California, Albert C. Martin, 1926–27. Distant view, with golf course in foreground.

In the Johnson project, however, the diagonal geometry does more than mimic the surrounding landscape's shapes. It takes on a dynamic role in physically directing the beholder to the larger aspects of the site and environment. The location of the entrance to the property at the upper end of the causeway (in the lower right-hand corner of the perspective) would have forced one to ascend the canyon to a point beyond the house itself before turning left, 30 degrees, to traverse the long diagonal leading to the main entrance of the house. In its pivoting action, the diagonal causeway directs the view down the length of the canyon to the peak of the Panamint Mountains across the valley, while the house itself picks up that motion to follow the spur of land on which it sits.

Johnson rejected the Wright design as too primitive, too much like "an adobe Indian village" rather than "a real hacienda, which we wanted."[8] In late 1925 he looked to others for help, and by 1931 a team of designers produced the proto-Disney, Spanish Colonial fantasy that was called Scotty's Castle and that soon became the must-see tourist destination of the National Monument. Whether Wright's design, had it been realized, would have attracted the same number of tourists is not easy to determine. What one can say for sure, however, is that it would have been a very early and extremely powerful expression of a modern desert architecture.

While Scotty's Castle was under construction, the first road was built in the Valley and two hotels were opened. The rather simple Stove Pipe Wells Hotel, completed in November 1926, was followed by the luxury Furnace Creek Inn, designed by the socially prominent Los Angeles architect Albert C. Martin and opened three months later. Like Scotty's Castle, it too adopted the Spanish Colonial style for its architecture, although it was much closer to Wright's site planning in terracing down the hill slope to afford extensive views of the distant landscape to its well-heeled guests. In its focus on the provision of views as well as its inclusion of a golf course as a major outdoor activity, the Furnace Creek Inn epitomized the new type of luxury desert resort of the Southwest.

Arizona Biltmore Hotel, Phoenix, Arizona, Albert Chase McArthur, 1927–29. Children preparing to go horseback riding in front of hotel, early 1930s.

PHOENIX AND CHANDLER, ARIZONA: ARIZONA BILTMORE AND SAN MARCOS-IN-THE-DESERT HOTELS

Almost exactly contemporary with the Furnace Creek Inn was the even more luxurious Arizona Biltmore Hotel in the desert on the outskirts of Phoenix. Designed in 1927–28 and opened in early 1929, it was intended to capitalize on the city's redirection from being a place mainly for health seekers to one devoted to the winter tourist trade of wealthy Easterners and Midwesterners. It was the brainchild of the three McArthur brothers, transplanted Chicagoans who had grown up in a house designed by Wright in the 1890s.[9] One of the brothers, Albert Chase McArthur, was an architect, and he designed the hotel. Having worked for Wright twenty years earlier in the his Oak Park office, and planning to use his former employer's system of reinforced concrete-block construction, known as textile-block, McArthur asked Wright to consult on the design. With no other jobs in sight at the time, Wright offered to come to Phoenix, which he did in January 1928, staying through April. There has been much debate on Wright's role—no doubt quite minimal—in the design of the Arizona Biltmore, but what is more important for us is that this initial exposure to the Arizona desert brought him a commission of his own that, though never built, would become a landmark in desert resort architecture and provide an experience of living in the desert that changed the way Wright managed his later career.[10]

The Arizona Biltmore provided opportunities for its guests to hike in the mountains behind the hotel as well as to go horseback riding and take part in "cowboy barbecues" in the nearby desert, but the hotel had from its beginning a sheltered and urbane character that focused on the more social and genteel outdoor activities of tennis, golf, and swimming. These attracted the "rich and famous," including many Hollywood celebrities. The hotel also set aside two-thirds of its original 600 acres for development, and

San Marcos-in-the-Desert Hotel project, Phoenix South (formerly Salt River) Mountains, Phoenix, Arizona, Wright, 1928–29.
Aerial perspective.

scores of its clients, such as Kirk Douglas, Nelson Rockefeller, and Henry and Claire Booth Luce, built private houses on the property to serve as winter residences. As a quasi-urban, activity-oriented resort with a firm basis in the economics of large-scale real estate development, the Arizona Biltmore proved to be a lasting success, but one that had little relationship with the kind of hotel Wright was to design for the area while acting as a consultant to the McArthurs.

In the early months of 1928, while in Phoenix, Wright received the commission for the San Marcos-in-the-Desert Hotel, which was intended to be an entirely different sort of operation. The client, Alexander Chandler, had traded a career as a veterinary

San Marcos-in-the-Desert Hotel project. Section through guest room wing (redrawn c. 1940).

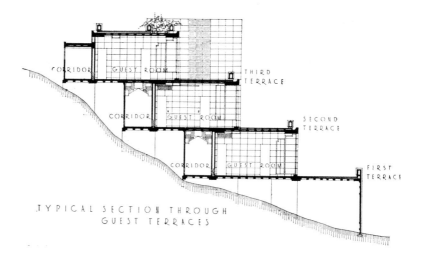

San Marcos-in-the-Desert Hotel project. Panoramic photo of view from building site, looking south, 1928.

surgeon for one in the irrigation, land reclamation, agribusiness, and winter tourism fields, through which he accumulated vast holdings of land along with money and connections. In 1912 he established a town, which he named after himself, 23 miles southeast of Phoenix, and built a resort hotel on the main square, which he called the San Marcos, after Fray Marcos de Niza, the first missionary to arrive in Arizona. Not content with this in-town, rather unexciting location, Chandler acquired, between 1924 and 1926, 1,800 acres of remote, unspoiled desert on the southern slopes of the Salt River Mountains (now called the Phoenix South Mountains), where he planned to construct what he eventually told Wright should be "an undefiled 'desert resort' for wintering millionaires."[11] There was to be no golf course, no tennis courts, no swimming pool, no manicured lawns—just the architecture and the desert in direct communication with one another—or at least that is how Wright described and interpreted the brief.[12]

The general area where the hotel was to be was in fact so remote that it was entirely undeveloped until a little over twenty years ago, and the actual site is still reachable only by a tortuous, unmarked trail. Wright completed the preliminary designs in 1928 and finished the working drawings and staked out the buildings on the site in 1929, expecting construction to begin the following year; but the stock market crash intervened and put an end to any thought of that. The 110-room south-facing hotel was designed as a series of stepped-back terraces forming two wings branching out from a central block. The wings follow the contours of the hill slopes, angling out on the left into a 30-60 degree triangle. The setbacks of the terraces translate the 30-60 degree geometry of the plan into section so as to provide each guest suite full exposure to the warm winter sun as well as extensive views out over the desert landscape to the south.

A panoramic photograph Wright created to study the design gives a good sense of the landscape looking south. The photograph also shows the dry wash running from right to far left that comes out of the gorge that cuts through the mountain ledge and splits the building in half. Wright planned to use the wash for the entrance drive of the

San Marcos-in-the-Desert Hotel project. Concrete block model, 1929.

hotel, making only minimal changes to its surface. The central block, housing the lobby and reception area on the lower floor and main dining room above, bridges the entrance drive and provides space for a turnaround in a small rear court. Everything was done, in effect, to minimize the building's intrusion into its landscape setting while maximizing the guest's appreciation of the setting's natural character and beauty. Wright stated that he "did not want to dig into the water-washed surface of the Desert," and the guest wing section reveals how he intended to eliminate any deep foundations and use the rocky ground itself as the basis for minimal footings.[13]

The aggregate for the concrete blocks, a scale model of which Wright built in 1929, was to have been drawn from the hillside so as to make the structure appear at one with the landscape. This symbiosis was not merely with the topography but extended to the plant life that is one of the unique features of the Sonoran desert. The blocks are articulated by vertical pleats, notched every now and then to imitate the appearance of the fluted surface with stippled arris of the surrounding sahuaro cactus. Wright claimed that the ribbed structure of the sahuaro provided a model for his system of reinforced concrete construction as well as a perfect example of how desert flora and fauna deal with the problem of the ever-present sun. "Sun-acceptance by way of pattern is a condition of survival" in the desert, he wrote, and such "sun-acceptance in building means that [the] dotted-line in outline and wall surfaces that eagerly take the light and play with it and break it up and render it harmless or drink it in until sunlight blends the building into place with the creation around it."[14]

Wright's remarkable and very intimate understanding of desert conditions was in large part due to the fact that while developing the final plans for San Marcos-in-the-Desert from January to May 1929, he chose to live in the desert about a mile south of the hotel site in a camp he designed to accommodate his family and office staff. Named Ocatilla after the spindly, candle-flame cactus that abounded at the site, the temporary

Ocatilla (Wright Camp), Phoenix South (formerly Salt River) Mountains, Wright, 1929 (demolished). Interior court with concrete block model for San Marcos project.

camp was composed of a series of canvas-roofed boxboard bases connected to one another by a zigzagging wall that formed a courtyard just below the crown of the low mound along the contours of which the structures were disposed.[15] The 30-60 degree geometry of the planning and the asymmetrical placement of buildings were intended to echo the shapes of the desert landscape. The wood was painted a rose color to blend with the desert floor and to tie in with the scarlet red color of the triangles at the ends of the white canvas roofs that was meant to resonate with the red flower of the candle-flame cactus. The scale model of the San Marcos blocks occupied the upper end of the courtyard just to the left of the drafting room.

TALIESIN WEST

Although Wright assumed that he would reuse Ocatilla as the site office for the hotel when it went into construction the following year, the camp was always meant to be temporary. Its publication in Europe and America, however, preserved for it a certain permanence in the growing modern interest in nomadic forms.[16] For Wright himself, its permanence as a response to the issue of how to build in the desert was secured by the seasonal home and headquarters for the Taliesin Fellowship that he began construct-

Taliesin West. Aerial view, looking east, 1949.

Taliesin West. Terrace prow under construction, with drafting room to left, ancient petroglyph boulder between it and dining room, and Sawik Mountain in distance, late 1939 or early 1940.

ing in 1938. As noted, Taliesin West was built on unincorporated land 17 miles east of Phoenix, well beyond any urban settlement at the time. While not as remote as Ocatilla or the San Marcos site, the land Wright purchased had no water, no electricity, and no road access, paved or otherwise.[17]

Wright located the camp, as he referred to the place, at the base of the McDowell Mountains, on relatively flat land between two washes. Following the model of Ocatilla, he planned a grouping of separate, interconnected units, each to be roofed by an angled structure of canvas stretched over wood frames. But unlike Ocatilla, where the base of the separate units was wood, here Wright sought a more lasting solution by incorporating the colorful quartzite boulders found lying around the site into forms filled with concrete to create what he referred to as a "desert rubble masonry."[18] This same conglomerate was then used to form raised terraces, platforms, and walkways that linked the various units to one another as well as to the natural and man-made landmarks of the surrounding desert landscape. Ancient Indian petroglyph boulders discovered at the site, such as the one on top of the steps in front of the pictured drafting room, were

Taliesin West. Garden, or living, room of Wright living quarters, 1947 (photograph Ezra Stoller © Esto).

placed in significant spots to reinforce a link to the cultural as well as topographical history of the site. But Wright never wanted the sense of permanence to overpower or undermine the character of fragility and transience of human existence in the desert, and so he kept his own living quarters more or less as a temporary-looking affair and required the incoming student apprentices to live for at least a season in tents in the desert behind the buildings of the camp.

WRIGHT AND ARIZONA

During the twenty years that Wright made Arizona his winter residence and office, he designed only one other project for the state's burgeoning tourist industry, a visitor center and observation tower for the Meteor Crater site in the northern reaches of the Sonoran desert that was never built (1948).[19] On the other hand, Wright himself and the buildings of Taliesin West became a major tourist attraction that has grown exponentially over time. *Arizona Highways*, the boosterish, tourist-oriented publication of the state's Highway Department, featured Wright's work on several occasions and devoted major spreads to Taliesin West in 1940, 1949, and 1956.[20] In recent years, as Greater Phoenix has become the fastest-growing metropolitan area in the United States, visitation figures at Taliesin West have reached 125,000 annually, making the site one of the leading attractions for cultural tourism in what is now commonly referred to as the Valley of the Sun. And the tourists come to Wright's buildings not only for cultural reasons but also as part of the new ecotourism. Desert walks, featuring explanations of the flora and fauna of the Sonoran desert to the visitor, have become a popular option at the site.

All of this tourist activity at Taliesin West did not come out of nowhere. Wright's buildings today, as compared to when they were first occupied by the seventy-year-old architect and his family and band of young student-apprentices, represent a utopian oasis in an almost absent desert, now serving more as a cautionary tale than as an example of how to live in concert with the special conditions imposed by the desert environment. Wright was fully aware of the interrelationship between tourism and development, although he found himself helpless to do anything about it.

Soon after completing the design of San Marcos-in-the-Desert, he wrote that the desert around Phoenix "will make the playground for these United States someday."[21] And shortly after completing the first phase of building operations at Taliesin West, he wrote in *Arizona Highways* a plea for the preservation of natural resources over man-made development, calling for "less salesmanship" and "less building" and an upscale approach to the provision of tourist facilities: "It is the grandeur of this great desert-garden that is Arizona's chief asset.... [T]he American people will come ... on account of the desert but must find Arizona a good host. They will come to get away from 'weather:' come to play and some of them come to stay in the invigorating sunshine.... [But] all of Arizona is not a large enough playground for the United States. So it may well remain caviar to the average and yet prosper mightily."[22]

For Wright then, this desert was to be preserved for a rather elite audience. "I, for one," he continued, "dread to see this incomparable nature garden marred, eventually spoiled by candy-makers, cactus hunters, careless fire-builders and the fancy period-house builders, as well as the Hopi Indian imitators, or imitation Mexican 'hut' builders. They will soon destroy the most accessible parts unless Arizona people have the sense to stop them." The "conservation of desert creation," he stated, "grows even more important. It is more important to government than the conservation of the forests and is important as any conservation of water in mountain streams. Should the legislature of the state neglect to protect its true reserves in this—eventually, yes, its greatest financial resources—the Federal Government should take Arizona over and protect it or future Arizona may yet live to curse its short sighted—too long-lived legislators."[23]

These are pretty strong words from an avowed decentralist. They can obviously also be discounted on the grounds of what we would today call NIMBYism. But they are worth considering in the context of the general theme of this volume and especially in the context of Wright's desert architecture, since they throw into relief the complex relationship of desert tourism to development in the first half of the twentieth century and the perhaps naive but certainly compelling attempt on the part of one of that century's most important architects to create an ideal, indeed utopian, space between those two poles.

NOTES

1   For the history of the area, see Bradford Luckingham, *Phoenix: The History of a Southwestern Metropolis* (Tucson: University of Arizona Press, 1989), especially 69–176.

2   See Keith L. Bryant, Jr., *History of the Atchison, Topeka and Santa Fe Railway* (New York: Macmillan, 1974); Diane H. Thomas, *The Southwestern Indian Detours: The Story of the Fred Harvey/Santa Fe Railway Experiment in "Detourism"* (Phoenix: Hunter Publishing, 1978); Marta Weigle and Barbara A. Babcock, eds., *The Great Southwest of the Fred Harvey Company and the Santa Fe Railway* (Phoenix: Heard Museum, 1996); and Kathleen L. Howard and Diana F. Pardue, *Inventing the Southwest: The Fred Harvey Company and Native American Art* (Flagstaff: Northland Publishing, 1996).

3   On Death Valley, see Richard E. Lingenfelter, *Death Valley and the Amargosa: A Land of Illusion* (Berkeley: University of California Press, 1986). On Johnson and Death Valley Scotty, see Hank Johnston, *Death Valley Scotty: The Man and the Myth* (Yosemite, CA: Flying Spur Press, 1974); and H. Johnston, *Death Valley Scotty: The Fastest Con in the West* (Corona del Mar, CA: Trans-Anglo Books, 1974).

4   Lingenfelter, *Death Valley*, 445–447.

5   Wright describes the trip in his *An Autobiography* (London, New York, and Toronto: Longmans, Green, 1932), 253. For a fuller documentation and analysis of the commission, see my *The Architecture of Frank Lloyd Wright* (Princeton, NJ: Princeton University Press, 1996), 173–189; as well as Dorothy Shally and William Bolton, *Scotty's Castle* (Yosemite, CA: Flying Spur Press, 1973); Robert L. Sweeney, *Wright in Hollywood: Visions of a New Architecture* (New York: Architectural History Foundation; Cambridge, MA, and London: MIT Press, 1994), 107–114; and David G. De Long, ed., *Frank Lloyd Wright: Designs for an American Landscape, 1922–1932* (New York: Harry N. Abrams, in association with Canadian Centre for Architecture, Library of Congress, and Frank Lloyd Wright Foundation, 1996), 66–80.

6   Wright, *Autobiography* (1932), 304.

7   Ibid., 305. For a fuller discussion of Wright's use of diagonal geometry, see my "Frank Lloyd Wright's Diagonal Planning Revisited," in Robert McCarter, ed., *On and By Frank Lloyd Wright: A Primer of Architectural Principles* (London: Phaidon Press, 2005), 232–263.

8   Johnston, *Death Valley Scotty: Fastest Con*, 104. Sweeney, *Wright in Hollywood*, 114, however, maintains that the reason was fundamentally financial.

9   See my *Architecture of Frank Lloyd Wright*, 197–198; Margaret Dudley Thomas, "The Arizona-Biltmore: The Queen of Internationally Honored Resort Hotels," *Arizona Highways* 50 (April 1974), 14–23; and Sweeney, *Wright in Hollywood*, 120–140.

10 On the controversy over Wright's role in the design,

see, aside from the works cited immediately above, "The Arizona-Biltmore Hotel, Phoenix, Arizona. Albert Chase McArthur, Architect," *Architectural Record* 66 (July 1929), 19–55; Letters-to-the-editor from Albert Chase McArthur and Frank Lloyd Wright, in "Behind the Record," *Architectural Record* 89 (June 1941), 7; Olgivanna Lloyd Wright and Bruce Brooks Pfeiffer, *The Arizona Biltmore: History and Guide* (Scottsdale, AZ: Frank Lloyd Wright Foundation, 1974); and Warren McArthur Jr., "The Arizona Biltmore, the McArthur Brothers, and Frank Lloyd Wright," *Triglyph: A Southwestern Journal of Architecture and Environmental Design* 6 (Summer 1988), 36–47.

11 Wright, *Autobiography* (1932), 301.

12 On the San Marcos-in-the-Desert Hotel, see Wright, *Autobiography* (1932), 301–303, 308–309; my *Architecture of Frank Lloyd Wright*, 197–201, 206–215; Sweeney, *Wright in Hollywood*, 140–169, 100–116; De Long, *Frank Lloyd Wright*, 100–116; and Reyner Banham, *Scenes in America Deserta* (Salt Lake City: Peregrine Smith Book, Gibbs M. Smith, 1982), 69–76.

13 F[rank] Ll[oyd] W[right] to A[lexander] J. Chandler, 24 July 1929, Frank Lloyd Wright Archives, Frank Lloyd Wright Foundation, Scottsdale, Arizona.

14 Wright, *Autobiography* (1932), 304; and Frank Lloyd Wright, "To Arizona," *Arizona Highways* 16 (May 1940), 11.

15 On Ocatilla, see my *Architecture of Frank Lloyd Wright*, 201–206; and Sweeney, *Wright in Hollywood*, 145–148. Wright's own description is in *Autobiography* (1932), 302–306.

16 "Desert Camp for Frank Lloyd Wright, Arizona. Frank Lloyd Wright, Architect," *Architectural Record* 68 (August 1930), 189–191; and Siegfried Scharfe, "Frank Lloyd Wright," *Baugilde* 13 (July 1931), 1164–1171.

17 See my *Architecture of Frank Lloyd Wright*, 254–297. For the architect's own description of the design and construction, see Frank Lloyd Wright, *An Autobiography*, new ed., rev. and enl. (New York: Duell, Sloan and Pearce, 1943), 452–455.

18 "Frank Lloyd Wright," *Architectural Forum* 88 (January 1948), 88.

19 The site was privately owned by the Burton Tremaines who later commissioned their friend Philip Johnson to do a new design, which was built in 1955.

20 Raymond Carlson, ed., "Mr. Frank Lloyd Wright, the Taliesin Fellowship, and Taliesin West," *Arizona Highways* 16 (May 1940), 4–15; R. Carlson, ed., "Frank Lloyd Wright and Taliesin West," *Arizona Highways* 25 (October 1949), 4–15; and R. Carlson, ed., "Frank Lloyd Wright and Taliesin West," *Arizona Highways* 32 (February 1956), 12–29. Wright designed a house for Raymond Carlson, the editor of the magazine, which was built in Phoenix in 1950.

21 Wright, *Autobiography* (1932), 303.

22 Wright, "To Arizona," 12, 8–9.

23 Ibid., 9, 12.

AZIZA CHAOUNI

4

THE KASBAH HOTEL: MYTHOPOESIS
IN THE MOROCCAN TOURIST LANDSCAPE

The architecture of the hotels in the south of Morocco is uniquely recognizable, like
Alpine cottages in their distinctive environment. The pre-Saharan area of Morocco,
encompassing the valleys of the Draa, Ziz, and Dades, is dominated by hotels emulat-
ing vernacular Kasbah architecture. Indeed, not a single hotel in this region does not in
some way refer to the typology of the Kasbah. A Kasbah, or *tighrem* in Berber, is a fortified castle located on the ridges of Morocco's pre-Saharan valleys, constructed from
unbaked loam and decorated with geometric motifs. It usually has two or more stories,
is centered on a small courtyard, and is characterized by a square plan and two or four
projecting corner towers that taper to a pyramid and serve to guard precious water
resources. The Kasbah was usually the homestead of an important feudal family and
their extended relatives, and it also served as a fortified farm with a communal granary.
It can stand alone, in clusters, or be part of a *ksar*, a fortified earth-built village.[1]

Two principal reasons underlie the dominance of the Kasbah as a referent for
hotels. First, the Kasbah, from the advent of tourism in southern Morocco in the 1940s,
has been the principal iconic image within the tourist *imaginaire*, inscribed in the discourses of consumption of tourists and hoteliers alike.[2] Second, the commodification
of the landscape in pre-Saharan Morocco has produced and marketed a one-dimension-
al view that has to be preserved and thus resists variations from the Kasbah vernacular model. The problem, however, with emphasizing the persistence and homogeneity
within the Kasbah typology is that it obscures ruptures within Morocco's history and
their effects on the built environment in the south. Indeed, it seems counterintuitive
to expect that the same imagery and iconography can be promoted by both the French

colonialist state, with its Orientalist and civilizing thrust, and the nationalizing postcolonial, independent Moroccan state, with its modernizing aspirations and rejection of its colonial past. Even the dominance of the centralizing postcolonial state underwent a major revision within the region in the 1980s, when the tourist industry was privatized and decentralized.

Why were there no alternatives to the Kasbah hotel typology? A possible course of action is to look below the surface of apparent similarity to understand not only what differences exist but how they could be accommodated. If the theme of the homogeneous approach is the "hegemonic image/discourse of the Kasbah," then the theme of the heterogeneous approach has to be "hybridization."[3] Indeed, once one looks more closely at the Kasbah hotels in the Moroccan pre-Sahara, one sees differences that include scale, material, integration of program, and relationship to landscape. The heterogeneous approach to representing the built environment of pre-Saharan tourism emphasizes the ability of the "Kasbah" to morph and accommodate various demands. As a result, the Kasbah becomes a reinvested signifier, able to cover and legitimize hotel programs under an illusory reference to tradition.

Using those two approaches to analyzing the Kasbah hotel typology, on both an architectural and a historical level, this essay will attempt to unveil the mechanism by which the Kasbah referent succeeds at simultaneously being malleable and the purveyor of a continuous iconography.

COLONIAL ERA: THE KASBAH HOTEL AND "BERBERISANCE"

The first hotels built in the Moroccan south after its conquest in 1934 were developed by a private steamship company, the Companie Générale Transatlantique (CGT), which wished to extend its network of all-inclusive tours. The hotels were labeled *gîtes d'étape* (rest houses) and erected along a dirt road punctuated by garrison towns and beautiful oases: Erfoud, Boumalne, Tinerhir, Zagora, and Tafraout. In the absence of a referent for building in a desert setting, the CGT turned toward the vernacular architecture model, the Kasbah, as a formal base for its hotel accommodations.[4] This choice was motivated as much by the company's travel dictum, which promised "a unique opportunity to taste the opaque and secretive intimacy of the southern Berber culture," as by the political orientations of French colonial discourse in Morocco.[5] Indeed, Morocco's colonial policies provided a nurturing ground for the adoption of local styles. The colonial architectural landscape in Morocco was stamped by the domineering preservationist vision of General Lyautey, which promoted a politics of aesthetic assimilation and built heritage preservation and documentation—a politics that culminated in the *Arabisance* style, a term first coined by architectural historian Francois Beguin as the adaptation of indigenous Arab architectural forms to French colonial buildings.[6]

Thus the desert hotels of the CGT portrayed the first instance of what we might call *Berberisance* in Morocco, in line with Lyautey's aspirations. This new form of hybridization was marked by exclusive ambivalence. In fact, in all CGT hotels, some elements from the vocabulary of the Kasbah are juxtaposed with International Style architec-

Collage of photos of Kasbah hotels

Photo and typical Kasbah plans and sections (drawings by the author)

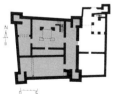

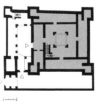

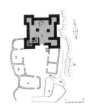

Advertising photo from Ministry of Tourism brochure, 1967

Cover of brochure of the Companie Générale Transatlantique (undated)

tonic components, in a balanced manner. For instance, if the introverted, square plan of the Kasbahs is elongated and flanked by both a viewing loggia and an off-centered tower, the massing and height remain close to those of the Kasbah. This volumetric similitude is furthered by the hotel's localization on a high ridge overlooking the valley, analogous to the Kasbah's strategic location. Similarly, if the main lobby and restaurant use local furnishing and construction materials, cooks and food are typically Parisian and the rooms have state-of-the-art bathrooms and bedding. Finally, although built in concrete, the hotels have a reddish finish close to the color of the Kasbah's earth material. This ambivalent approach to the appropriation of the Kasbah echoes the colonial discourse dialectic that both elevated indigenous culture as exotic and attractive and denounced it as inferior and in need of civilizing. It should be noted that colonial tourism, the defining force behind those hotels, was part of the colonial civil duty euphemized as the "mission civilisatrice." Colonial tourism was presented as a vehicle for education through first-hand experience about the facts of colonialism and the range of mutual "benefits" for both metropolitan and colonial subjects.[7] In this context, the Kasbah hotel functioned as a "viewing platform" of the untamed, raw Moroccan South, gradually marked by France's modernizing imprint.

INDEPENDENT MOROCCO: THE KASBAH HOTEL AND CRITICAL MODERNISM

After Morocco's independence in 1956, the five *gîtes d'étapes* of the CGT remained the only accommodations available in the Moroccan south until the mid-1960s. At that moment the newly independent Moroccan state, coming out of its first financial crisis, selected tourism as its most important sector for investment after agriculture, to spur economic development.[8] In its development plans in the 1960s and 1970s, the Moroccan government invested about 7 percent of its budget in tourist infrastructure, especially targeting its poorest regions, which were not considered part of the "utilitarian Morocco": the Mediterranean shore and the southern provinces, in Boumalne, El Kelaa, and Taliouine. From 1964, the Moroccan state set in motion a politics of economic liberalism coupled with a modernization campaign. In official reports, the objective behind developing tourism in those rural southern regions is "not only to create rural middle classes with new consumption habits, but also to change the traditional behavior of the population judged too static, and consequently, less open towards the new exigencies of modernization."[9] Despite such drastic modernizing goals buttressed by a state-sponsored institutional architecture built by the innovative "Rabat School"—a group of architects perpetuating the legacy of the CIAM Morocco (GAMMA) and known for their impressive modernist reconstruction of Agadir[10]—the Moroccan government chose the Kasbah as a referent for its southern hotels. This retreat from an *avant-gardist* vocabulary symbolizing progress and return to a direct relationship with a vernacular model can be linked to two main contextual causes. First, strong nationalist discourse encouraged the promotion of a regional architectural style to strengthen patriotism and nation building. Second, the "Rabat School" architects tended to embrace not only traditional materials and techniques but also vernacular typologies. In fact, these architects were influenced not only by the mounting critical discourse of Team 10, whose members advocated a more situated modernism, but also by the legacy of one of GAMMA's founders, George Candilis, [11] who integrated the courtyard typology and local climate constraints into two experimental housing project blocks in Casablanca's Carrières Centrales (Nid D'Abeille and Semiramis).[12]

Going a step further, new Kasbah hotels used a definitely modern aesthetic with a traditional referent. The two architects who designed the series of state-sponsored southern Hotels Faraoui and Demazières (FDM) transposed onto the Kasbah referent principles from the "Rabat School" that advocated the adaptation of CIAM's precepts to Morocco's milieu. This hybridization strategy is one of analytical reinterpretation. The program of the hotel, where staggered cubic rooms are organized around a central swimming pool and terrace, is conceived to evoke the general massing and introversion of an aggregation of Kasbahs rather than the direct reproduction of the referent. This choice is judicious as it allows 100-room hotels to relate on a one-to-one scale to the Kasbah and thus minimizes their visual impact on the landscape.

The interpenetration of historically disparate forms generates an innovative hybrid that starts delineating its own identity, almost obscuring its referent, the Kasbah, and its modernist legacy. By requalifying spatially the built environment and generating a symbiotic language, these hotels could fit into Liane Lefaivre and Alexander Tzonis's definition of successful critical regionalism, which advocates the defamiliarization of regional elements.

BACK TO TRADITION

In 1981, King Hassan II gave a speech to the Moroccan architectural association, urging members to embrace their "Moroccanness" and avoid the Westernized trends with which Moroccan architecture had been afflicted. This nationalistic discourse aimed at shaping and consolidating the Moroccan identity came after a decade that included two unsuccessful coups that challenged the monarchy's sovereignty and the country's unity. Shortly after this speech, most public commissions, followed by the private sector, adopted a regionalist style that varied according to geographical area, but had an arabo-andalousian accent throughout the country. By accreting bits of traditional architecture onto contemporary typologies without a clear formal agenda, this uncritical regionalist trend generated a postmodern kitsch language, a far cry from the creative aesthetic of the "Rabat School." Even in the far provinces of the south, regionalism dominated, yet buildings retained a strong local character. The Kasbah still remained the main referent for southern architecture, especially for hotels. The advent of "Kasbah regionalism" coincided with the first waves of mass tourism in southern Morocco, spurred by both the opening of the Ouarzazate international airport in 1986 and the booming of the film industry in the region. This growth in clientele implied not only a drastic increase in the size of hotels but also an upgrade and diversification of amenities in the form of spas, gyms, and conference centers. The average size of four- and five-star hotels built in that period was 200 rooms. Accordingly, the Kasbah typology required a steroid infusion to meet those new requirements. As a result, the proportions of the Kasbah were blown out of scale: from an average of 15 x 15 meters, it was multiplied by a factor of three or sometimes four. Hence this hybridization strategy relied on scalar manipulations.

Despite the significant increase in size, the main spatial hierarchy of the Kasbah was kept intact, as it perfectly corresponded to the gated hotel typology, the preferred model of mass tourism. These gigantic Kasbah hotels are encircled by high walls and corner towers, are introverted toward a central swimming pool (often of Olympic dimensions), and have access to the exterior landscape only through terraces.

Kasbah regionalism was also deployed in a miniaturized format for small hotels, usually owned by locals. In those cases, the Kasbah's architectonic components are minimized: large entrance gates become doors, four- to five-story corner towers are reduced to pilasters. The plans of those miniature Kasbahs are either courtyard centered (enhancing the Lilliputian effect) or L- or U-shaped, but the main façade facing the road remains a direct referent to the Kasbah. Similarly, in super-Kasbahs, the referent seems to hold a stronger presence on the façades, in compliance with the production of the cultural landscape of the oasis as a "postcard" image ready for con-

CGT hotel in Zagora (undated)

Photo of Hotel Boulemane by Faraoui and Demazières (undated)

sumption. In contrast, the interior spaces display an array of transnational influences. Interiors are cluttered with a wide spectrum of deterritorialized elements, not only from other Moroccan provinces (such as green tiles and horseshoe arches) but also from far-away inspirations such as Hawaiian beach huts, Alpine stone fireplaces, African wood-work, and Hollywood memorabilia. As a result, the reappropriation process of the Kasbah in this case results in what Lefaivre and Tzonis have called a "romantico-commercial regionalism," where the building is reduced to an object of hedonistic consumption.

THE ECO-KASBAH HOTEL

Amid the hegemonic Kasbah regionalist style, a current of preservationism has emerged in the past decade, coinciding with the rising popularity of an alternative mode of tour-ism that moves beyond the usual paradigm of acculturation and development: eco-tourism. A facet of the larger global green movement, ecotourism aims to touch places and traditions as lightly as possible and sustain local ways of life by promoting simple accommodations built of traditional materials. Today, the growing eco-clientele longs to stay in authentic mud Kasbahs. To respond to this new demand, eight Kasbahs have been restored in southern Morocco and turned into guesthouses since 1999. About fif-teen were built from scratch using traditional techniques and materials. To help these endeavors, a manual on restoration methodologies for earth architecture has been issued by UNESCO. In both replication and rehabilitation instances, the reappropriation of the Kasbah is articulated through a reprogramming hybridization process. Indeed, this new generation of Kasbah hotels aims to be as close as possible to their original ref-

Hotel Kella Mgouna by Faraoui and Demazières (undated)

Hotel in Mezourga in 2007

erent: they keep their plan, materials, construction techniques, and massing as close as possible to the traditional Kasbah, and adjust the hotel program to suit the form of their referent. Rooms are usually kept as guest rooms, first-floor communal and storage areas are converted into restaurants, and living rooms and detached buildings that housed relatives accommodate staff and services. However, modern amenities such as running water, showers, and toilets are incorporated, not without endangering the fragile foundations of the building. One of the advantages of this Kasbah hotel typology is the salvation of an ethereal architecture and its mode of construction that otherwise would have disappeared. Also, this typology represents the most sustainable hotel model because it minimizes energy use. The intellectual finality of the end product—a recycled receptacle or replica—remains questionable. Indeed, the claim for authenticity of these hotels is challenged by their rupture with the sociocultural context of the Kasbah. Moreover, their faithful reconstruction of a vernacular model lacks innovations and limits itself to a rigid retrofit.

In addition to this fundamental problem, some eco-Kasbahs have been seduced by thematization. For example, in the case of one of the oldest eco-Kasbahs, the Baha Baha in Nkob, rehabilitative reprogramming was applied in a first stage to a family-owned Kasbah. Then, in a second stage, a pool, outdoor huts, nomadic tents, and a *petanque* field were added. A third stage envisions a museum composed of a miniaturized oasis, and Kasbah and Berber dwellings. The owner explained that he wished to attract clients beyond the ecotourists. One wonders if the economic reality and the volatility of tourism trends will push the other eco-Kasbah hotels to become mini theme parks as well, where the Kasbah referent becomes one objectified vernacular prototype among others.

One could finally wonder if a similar fate awaits the Kasbah hotels from other generations. Rethinking the Kasbah hotels is even more urgent today, as they represent a large part of the Grand Sud's remaining "built heritage," thus becoming themselves a referent for their built environment.

NOTES

1   For definition of the Kasbahs, see Robert Montagne, *Villages et kasbas berbères: tableau de la vie sociale des Berbères sédentaires dans le sud du Maroc* (F. Alcan, 1930).

2   Revues du Cycling Tour de France.

3   Precisely because hybridization provides an avenue for alternatives and accommodation of various sociopolitical circumstances beneath a seeming orthodoxy.

4   Marthe Barbance, *Histoire de la Compagnie Générale Transatlantique: un siècle d'exploitation maritime* (Paris: Arts et Métiers Graphiques, 1955).

5   CGT 1936 brochure, Archives de la CGT, Le Havre, France.

6   Francois Beguin, *Arabisances: Decor architectural et trace urbain en Afrique du Nord 1830–1950* (Paris: Dunod, 1983), 69.

7   Ellen Furlough,. "Tourism, Empire, and the Nation in Interwar France," *French Historical Studies*, vol. 25, no. 3 (2002), 441–473, 443.

8   Mimoun Hilali, *Le developpement du tourisme mediterranean en harmonie avec l'environement. Le Cas du Maroc* (Rabat: Ministere du tourisme divisions des amenagements et equipements touristiques), 3.

9   Ibid., 82.

10 The Rabat School comprises: Jean François Zevaco, Patrice Demazières, Abdelslam Faraoui, Henri Tastemain, Ellen Castelneau, Riou, Pierre Mas, and Elie Azagury. See Thierry Nadau, "La reconstruction d'Agadir, ou le destin de l'architecture moderne au Maroc," in Maurice Culot and Jean-Marie Thiveaud, eds., *Architectures Francaises d'Outre Mer* (Liège: Mardaga, 1992), 160–172.

11 Jean-Louis Cohen, "The Moroccan Group and the Theme of Habitat," *Rassegna,* no. 52, (December 1992), 58–67.

12 Shadrach Woods, *Candilis-Josic-Woods: A Decade of Architecture and Urbanism* (Stuttgart: Karl Kramer Verlag, 1968).

VIRGINIE PICON-LEFEBVRE

5

FROM WAR TO PEACE,

THE MAKING OF A LANDSCAPE

This chapter focuses on the construction of the landscape as seen from the roads built by the French during their conquest of the Moroccan south against the Berbers, which eventually became tourist roads. It is about the way this landscape was described from those roads as a military object of surveillance and control, to become the typical view of southern Moroccan landscapes, the sight that tourists now expect. The hypothesis is that the tourist gaze parallels the military gaze—tourists and armies looking for the same things, but for different reasons. They are both looking for the best view without having to understand the culture, and avoiding any contact with the inhabitants. An additional effect is that tourism has transformed the relation of the inhabitants to their own territory.

Historically, tourism in southern Morocco was not defined through fixed destinations as was the north of the country (which was encountered through visits to imperial cities), but as travel through the landscape—to cross the Atlas mountains, to drive in the desert, to stay in the oasis. Tourism can be seen as the reenactment of the military movements of colonization, as well as the reenactment of the nomadic experience. At the same time, the movement of tourists has replaced the nomads' slow movement with a fast pace, with few interactions with the locals. The tourists with their different subjectivity, their different behaviors, are becoming a new sort of nomad in a global society characterized by mobility.

However, the effect on the territory differs. If nomads were accepted by the locals and contained in a certain area, tourists go everywhere and maintain their positions by spending money. One of the consequences is that the inhabitants have abandoned the

A map showing North Africa as one territory, from Général Mangin, *Regards sur la France d'Afrique*

previous locations of their traditional *ksour* (fortified villages) to move near the new roads to cater to the needs of tourists, selling souvenirs as well as opening restaurants and hotels. The shifting of most social and economic activity to sites along the modern roads complicates the renovation of historic villages, as most are now uninhabited. They became monuments and lost their previous meanings; no one is able to maintain them, and they are slowly but surely disappearing. The most striking case is Ait Ben Adou, a UNESCO World Heritage site, which is a ghost town, as everybody now lives near the road on the other side of the river, where tourists arrive every day by car or bus and stay for a night in hotels built for them along the river.

One can say that the military conquest has been followed by a more peaceful one by tourists, building a new economy, bringing a more reliable source of revenue, while provoking a complete change of the location and form of human settlement as well as a new social hierarchy. Those who were able to cater to the needs of tourists and settle along the road gained an advantage over those who stayed in their historic location.

When one visits the southern part of Morocco on the edge of the Sahara, looking for the desert, two things are surprising. First, the road itself is an important element of the experience, as many people are walking along it, and public services as well as utilities such as cafés and gas stations are located directly on it. Those roads are strange places because there are both a means of circulation and public spaces, even without being in a densely constructed zone. One can find hotels, mosques, clinics, and schools directly on the road. Second, it is not easy to visit the oases and the historic villages, as they are located at a certain distance. The distance presents the picturesque vision, seen in advertising posters and brochures—the one that tourists are seeking.

My discussion will follow two types of narratives: those drawn from the description of the "pacification" by Général Mangin in 1921 and later in 1937 by a geographer, J. Célerier, and others from the first tourist guidebooks of Morocco.[1] Once the southern territories were "pacified," French companies and automobile tour associations—for example, the CGT (Compagnie Générale Transatlantique)—quickly planned organized tours in the south of Morocco.

The descriptions from Joseph Dalton Hooker are both scientific and artistic, a common attitude among early travelers; from *Journal of a Tour in Morocco and the Great Atlas* (1878)

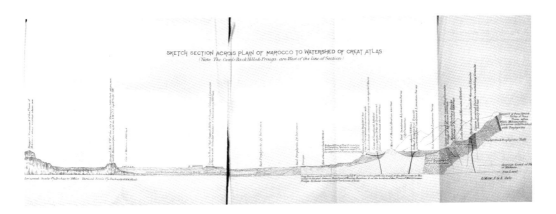

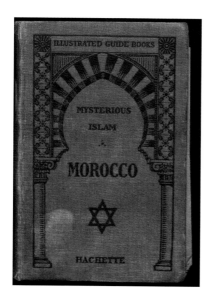
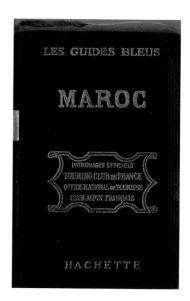

Front cover of the English and French versions of the Guide Bleu from 1929

THE MILITARY VISION OF THE LANDSCAPE

A beautiful landscape is of interest to the military as well as to the tourist. The geographer Yves Lacoste, born in Morocco and married to an anthropologist who studied Berber civilization, explains the link between the notion of the landscape and its description as the place of battle. He notices that, curiously, the definition of a beautiful landscape has common points with a landscape described by soldiers as favorable for fighting. One shared attribute of those two perspectives is concern for the view.[2]

A panoramic view makes it possible to control the movements of one's enemy as well as one's own army during battle. On the other hand, a beautiful panorama for a tourist must have parts visible and others hidden; it has to be open at certain points, closed in others—in short, it has to be diversified and full of surprises.

When Général Mangin explains the logic of the road construction in the south in his account in March 1929, *La conquête du Maroc,* he first describes the relationship between landscape and war. Morocco, he says, is related to the Atlantic side of Africa and presents three chains of mountains that separate it from Algeria and from the Sahara. That explains why the war in Morocco was different from Tunisia, for example, which is very accessible from the Mediterranean Sea, whereas Morocco's shore along the Atlantic is dangerous for sailors. Mangin's vision of southern Morocco is of an enclosed territory, easy to defend and difficult to attack, far from Western civilization.[3]

The military conquest of southern Morocco was therefore realized through the construction of roads to facilitate the movement of the army. The strategy was to create

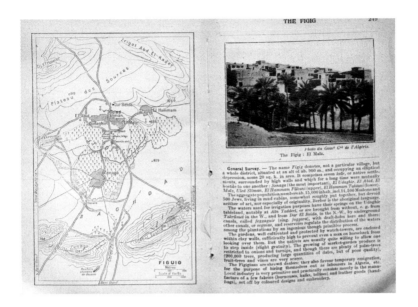

The description of the landscape as best seen from a distance, given the danger in coming closer;
Maroc, Guide Hachette (translated as *Mysterious Islam*), 1929

regional centers, from which the army could organize limited attacks for the purpose of pacification in every direction, with relatively few soldiers. When necessary, a larger attack could be organized with the entire army. From the regional center, the occupation/modernization of the country was then possible to impose. Each of those regional centers was under the command of a different general.

Célérier confirmed that the network of roads was organized from a main point to the periphery; for that reason, the different systems were not well connected.[4] The connections were better defined from north-south than east-west. The first road between the Atlantic and the pre-Sahara opened in 1917, going from Meknes Bou Denib through Tarzeft to Ksar es Souk. The roads were sited to facilitate control over the population. They were always built at a distance from existing villages, and the high points were occupied by military posts that were easy to defend.

The French occupation came from the north and moved south. If the military occupation was followed by the tourist one, it was for the army a way to maintain a constant form of permanent control along the road. Seen this way, the tourist presence assumes an ambiguous status of both invaders and providers. In Morocco this double perception may be more specific because of its history. But why have wealthy tourists accepted such a dangerous task? One may find the answer in the description of the landscape.

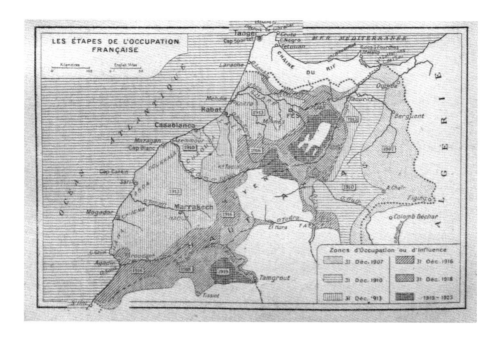

The Guide Bleu maps the progress of the colonization of Le Maroc

THE FIRST TOURISTS: WRITERS AND SCIENTISTS

It is common for a destination to be first described by a photographer, scientist, poet, or writer before being opened to mainstream tourist consumption. In the *Journal of a Tour in Morocco and the Great Atlas*, Joseph Dalton Hooker, director of the Royal Botanic Gardens of Kew, is both a tourist and a scientist. The book was published in 1878, the same year that the Baedeker guides for Palestine and Syria were released. We can define those years as the beginning of tourism in the Sahara. The introduction states: "Up to the date of our visit, the Great Atlas was little better known to geographers than it was in the time of Strabo and Pliny."[5]

Hooker described and represented the landscape before it became a tourist destination. The trip was very difficult to organize and required many servants and much material. Travel began in April 1871, and the first thing noticed was the climatic condition: "It is a pleasant thing to turn his face towards the South in the early part of the year, and to feel that [he] is about to exchange six or eight weeks of bitter easterly winds for the bright skies and soft breezes of the Mediterranean region."[6] Hooker learned from foreign residents that "Morocco is the China of the West ... [and] has remained more isolated and more impenetrable than even the Celestial Empire itself."[7] This observation identifies the search for mystery as one of the reasons to travel to such a destination. The lack of roads and the rugged mountain landscape described fifty years later by Mangin built on this idea, adding that it might be dangerous to travel in the region because of difficulties with the population.

When Hooker's group arrived in the mountains, they passed near several villages. "As a rule, the valley's tracks in the great atlas are carried on one side and do not approach near to the houses," he wrote. These positions were chosen to keep nomads at a distance while they were traveling along the tracks. One of the images they sketched is a panorama of the great Atlas from the city of Marrakech. Among the notable elements were the openness of the landscape with the walls of Marrakech, a tent camp, and the absence of roads. The representation of the landscape has several directions and no focal point, which can seem disconcerting when compared to the European landscape organized along a hierarchy of roads. Through Mangin et Célérier, the vision of the landscape from the road began to take shape, looking to the picturesque via the modern infrastructure.

THE GUIDEBOOKS' TOURISTS

In *La roue et le stylo*, C. Bertho-Lavenir describes the evolution of tourism after the diffusion of bikes and cars around 1895.[8] Bicycles and cars tourism would change the idea of travel in two ways. First, it would provide the freedom to visit unknown places without having to follow the railway schedule; and second, it would encourage the establishment of travel associations, such as the Touring Club de France, which were important for the economy and the collective experience of tourism. Travel associations played an important role by describing itineraries and providing road signs as well as guidebooks

The landscape as seen from the road today

and maps. They sought road improvements and the creation of new roads to give access to picturesque scenery. The associations also invented a new kind of tourism: touring by car. They also promoted the creation of tourist information offices, supported the protection of historic monuments (as possible destinations), and established guidelines for the standardization of hotels and inns. They published magazines to advertise new destinations. Morocco missed the first stage of the development of tourism, which happened in parallel with the railways, instead immediately becoming a destination for car tourism.

It is not surprising then that the Compagnie Générale Transatlantique, known at the French Line, offered tours by car in Morocco as soon as the country was open to tourism. Its director, Dal Paiz, built hotels along the road; he understood the potential of the desert tour and used his company lines to provide access to the desert through Algiers and Casablanca. He used the same strategy to find passengers as did the railway companies when they worked with Thomas Cook, the inventor of the package tour, as a way to increase the number of travelers.

But organized tours were not the only way to travel through the desert—visitors could also use guidebooks. The first guidebook for Morocco was published by Hachette in both French and English, but under different titles. In English the title refers to the mysterious Islam, while in French it relates directly to the country: *Le Maroc*.[9] The guide was published under the sponsorship of the Touring Club de France. It was first printed in 1921, with a foreword by the military governor, Général Lyautey, and described the transformation of the infrastructure of the country. The guide was published ten years after the signature of the protectorate and was a success, going through three editions by 1929. In both guides, in French and English, there is a map of the French occupation; that those territories are open to tourism is evidently assumed by the writer.

The road as the new spine for urbanization

The description of the desert established the idea of tension between modernity and the gentle and slow pastoral and town life of Moroccans. The landscape is alternately described as inviting or thoroughly wild, but the "roads are straight and easy to follow." The tours described by the guide are different from those offered today, since Algeria was a French colony. It describes the *Grand circuit Nord Africain*, from Tunis to Casablanca, a trip taking two to three months. There was also a shorter tour, from Casablanca to Marrakech and Mogador and back to Rabat in two weeks. The south is represented trough *La tournée du sud Saharien* that included Figuig and the Ziz valley.[10] Those tours are no longer possible today, as the relations between those countries are unstable and the frontiers closed.

The tours followed the routes opened by the French occupation in the Sahara, the guide explains: "The military services laid out tracks in every direction; in dry weather these tracks make fair running for motor cars." Some tours such as Figuig, however, offered train travel to reach hotels, including one built by the CGT in 1921.[11]

How did the guide define the specificities of the desert landscape? The Ziz valley illustrates well how the tourist arrivals immediately followed the military occupation. The Ziz was "pacified" in 1916, and in 1921 the first part of the Ziz was already open to tourists. Access roads were still rough, and there was no accommodation. The main feature of site was "its powerful charm; the palm grove in the *oued* [bed of a river] like a large green snake."[12]

The description gives the number of *ksour*, and their population, with the negative comment that villages are overpopulated. The battlefields are also described as one of the attractions: "Meski, where colonel Doury fought 20,000 men with a few battalions in 1916."[13] The visitor is invited to see, from a distance, "ksour and well-maintained fields,

New hotel and housing along the road (photos by author)

irrigated with care." There are no details about the architecture of the *ksour,* giving the impression that it was not useful to visit them but better to see them from a distance. The guide maintains the same attitude as those of the travelers of 1871. The hypothesis of a strong link between war and tourism may be confirmed through the rather naïve description the guide gives its readers.[14]

As one of the few early tourist destinations in the Sahara, the Ziz valley featured elements of the tourist landscape experience that would last until today: fascination with the lush palm grove, description from a safe distance of the *ksour,* and absence of contact with the population because of the constant feeling of possible danger—in other words, a fascinating place to look at, but not to stop in.

The modern roads used by tourists today connect high points of observation to former military centers, now booming little towns. The *ksour* and the oasis are still mainly seen as distant objects to be framed in a panorama, as in the Ziz valley. From the visitors' point of view, *ksours* are still not seen as welcoming places. This explains why there remains a gap between the new road and the old villages. At the beginning of this research, I thought that this gap was the result of the military origin of those roads. It seems, however, that this separation between the tracks and the villages is more ancient. Nomad tracks were located far from villages, and *ksour* stand at a distance from possible enemies or invaders.

The tourist landscape was constructed from the modern road as a place to look upon medieval Morocco. This focus probably explains the lack of interest in the conservation of *ksour,* as tourists were discouraged from visiting them—they became not a destination but an element of a picture. It is time now to reverse the direction of the gaze from the *ksour* to the road, from the oasis to the landscape, for the sake of more sustainable development and to be able to invent new programs for the renovation of these incredible places. If nothing is done, these landscapes will probably lose their specificity. It would be more interesting to abandon the nostalgic panorama of ruins and instead help the inhabitants modernize outdated structures, to better live in them and attract visitors not for superficial sightseeing but to experience a real and authentic place.

NOTES

1 Général Mangin, *Regards sur la France d'Afrique* (Paris: Plon, 1921).

2 Yves Lacoste, *Paysages politiques* (Paris: Le livre de poche, 1990).

3 Mangin, *Regards sur la France d'Afrique,* 77. Mangin includes a map of North Africa shown as one territory.

4 J Célérier, "La croisée des routes marocaines en Haute Mouloya," in *Mélanges de géographie et d'Orientalisme offerts à E F Gautier* (Tours: Arrault and Co., 1937), 146.

5 Joseph Dalton Hooker, *Journal of a Tour in Morocco and the Great Atlas* (London, 1878).

6 Ibid., 4.

7 Ibid., 30.

8 C. Bertho-Lavenir, *La roue et le stylo, comment nous sommes devenus touristes* (Paris: Le Champs médiologique Odile Jacob, 1999).

9 *Le Maroc,* Guide Hachette (Paris: Hachette, 1929), English translation entitled *Mysterious Islam,* Hachette Guide Book (Paris: Hachette, 1929), 30.

10 *Guide Hachette du Maroc* (Paris: Hachette, 1921), 50.

11 *Le Maroc.*

12 Ibid., 378.

13 Ibid., 400

14 The French edition of the guide has a map showing the phases of the French occupation, unlike the English edition. *Guide Hachette du Maroc,* 29.

ALESSANDRA PONTE

6

PHOTOGRAPHIC ENCOUNTERS
IN THE AMERICAN DESERT

The Indians do not like to be photographed.
—ABY WARBURG, Diary, May 3, 1896, Arizona[1]

Strictly speaking, one never understands anything from a photograph. [...] Today everything exists to end in a photograph.
—SUSAN SONTAG, *On Photography,* 1977[2]

The arid territories of the American Southwest have been the real (or fictional) theaters
of the mythical conquest of the West. The region is punctuated by magnificent pre-
Columbian ruins; and Native Americans represent a substantial portion of the population, live in reservations, and in some exceptional cases, such as the Pueblo Indians,
still occupy the lands of their ancestors. The desert landscapes have also been, and still
are, heavily used by scientists and the military to develop and test the most advanced
weapons. American Indians and war technologies have generated two significant and
apparently very distant forms of tourism. The first has a longer lineage, beginning at the
end of the nineteenth century. The second emerged in the early 1950s and is commonly
referred to as "atomic tourism." One may argue that in both cases, the objects of fascination and attraction are constituted by war and its effects. This is spectacularly clear
in relation to the phenomenon of atomic tourism. In the case of encounters with the
native inhabitants of the region, the history of past violence and the pain of present conflicts are less evident, if not hidden.

Paul Chaat Smith, a Comanche and an assistant curator at the National Museum of the American Indian, has written that in the United States, a most forgetful country, "whose state religion seems to be amnesia,"[3] Indian history needs to be relentlessly invoked. A significant portion of such history involves precisely the account of how Native Americans and their culture have been stereotyped and commodified to satisfy an ever-growing and variable tourist industry. One may say that tourism has been another form of conquest and subjugation, another Indian war. In such a war, as in previous ones, American Indians have valiantly engaged in developing forms of resistance to that quintessential tourist weapon, the camera.

TRAVELOG: 1997

I went to visit the American Southwest for the first time in 1997. I was already planning to write a book on the American desert and had read extensively on the topic, including books dealing with the Native American inhabitants. I knew about the pueblos of the Zuni, and the Hopi, and of the presence of Navajo and the other tribes living on reservations, and about the spectacular and mysterious pre-Columbian ruins. I was also aware of how, since the beginning of the twentieth century, the architecture, arts, and traditions of these peoples had been exploited, commercialized, and even transformed to serve the tourist industry. In addition, I was familiar with the ethnographic literature about the various tribes, from the notorious accounts of the Zuni written at the end of the nineteenth century by anthropology's first "participant observer," Frank Hamilton Cushing,[4] to the celebrated *Patterns of Culture* (1934), where Ruth Benedict established her famous opposition between the "Apollonian" cultures of the Pueblo of the Southwest and the "Dionysian" attitudes of the Native Americans of the Great Plains.[5] At the time, for almost two decades, the work of the first American ethnographers had been under intense critical scrutiny, as part of a general process of reassessment of the discipline. With the writings of Paul Rabinow, Edward Said, Roy Frank Ellen, James Clifford, George E. Marcus, and Michael M.J. Fisher,[6] anthropology's claims to provide authoritative interpretations and convey an authentic experience of other cultures had been radically challenged. The mirror had been turned, so to speak, on the discipline, revealing a rather disturbing picture. During the same period, from a similar perspective, tourism and tourists had been extensively investigated by sociologists, anthropologists, and experts of semiotics, all intent on demonstrating the hopelessly inauthentic character of the modern tourist experience.[7]

Before even arriving in the Southwest, I was therefore prepared to enjoy precisely the inauthentic nature of the experience and to accept the limitation of a role that I considered inescapable. I was going to be a tourist, consciously part of the global phenomenon of commodification of cultures. I had no illusions about the possibility of acquiring a superior or detached status by qualifying myself as a "traveler," "pilgrim," "observer," or "sympathetic researcher." This, I presume, was also the attitude of my companions. I was traveling with four other architectural critics and historians. None of us was American, and for all this was the first encounter with the region and its native inhabitants.

We landed in Albuquerque loaded with guidebooks and cameras. Each had at least one camera at the beginning of the trip and, before the journey was over, we had all acquired disposable cameras to take panoramic photos. We had the impression that panoramic photos best captured the arid and spectacular scenery. The truth is that no apparatus can really capture such landscapes. No matter how many commercials, films, photographs, or paintings by the best artists one has seen, no matter how much one has the feeling of already knowing these places, the reality of them is going to surprise, enchant, and overwhelm. Nevertheless, like all good tourists, we took hundreds of slides and photos, and bought postcards, more guides, more books, and more slides on sale at various tourist locations, not to mention every possible kind of souvenir, includ-

Desert landscape with tourists (friends of the author), American Southwest

ing Stetson hats, bolo ties, Navajo, Zuni, and Hopi jewelry, sand paintings, and katchina dolls. I don't think we missed a single tourist shop from Albuquerque to the Grand Canyon and back.

The airport of Albuquerque fully satisfied the theme-park experience we were expecting: fake adobe interiors, shops selling miniature sand paintings, dream catchers, and katchina dolls, together with restaurants serving Spanish rice and Texan fajitas. I am writing from memory (I didn't take notes during the trip), and what I remember next is the drive to Santa Fe with a detour to visit Bandelier pre-Columbian ruins, haunting and inscrutable in the freezing, transparent winter afternoon, and a first night in a very cold and uncomfortable Best Western Hotel.

Santa Fe: Two publications, bought during the trip and dated 1997, evoke part of the feeling of walking the streets, visiting the museums, and shopping around the plaza. The first, written by Chris Wilson, a professor at the University of New Mexico who lives in Albuquerque and is interested in architecture and the politics of culture, investigates the invention of the myth of Santa Fe and the "creation" of a "modern regional tradition."[8] Wilson's book meticulously maps the history of the occupation of the area, beginning with the so-called Pueblo Indian—settled people practicing agriculture—followed by the arrival of the nomadic ancestor of present-day Apaches and Navajos, and then by waves of Spanish and Anglo-Saxon colonization. After sketching a narrative of conflict, repression, and domination, but also of exchange and racial mixing, Wilson proceeds to demonstrate how, from the early 1900s, the city was deliberately designed to appear a romantic and exotic destination where three distinct and equally "picturesque" ethnic groups were happily living together in harmonious segregation.

The second book, the catalog of an exhibition, presented the systematic marketing of the entire region under the title *Inventing the Southwest: The Fred Harvey Company and Native American Art*.[9] An article about the show, published in the *New York Times*

Taos Pueblo (our car in front of the entry point), New Mexico

in December 1997, remarked on how Fred Harvey, an English immigrant, set the standard for masterful packaging in 1876. The Fred Harvey Company, in association with the Atchison, Topeka, and Santa Fe Railroad, operated dining cars and created along the line restaurants and tourist hotels designed in a style mimicking the adobe construction of the Spanish and Pueblo settlements. The company was also responsible for collecting, displaying, and organizing the sale of Indian antique and contemporary artifacts, from Navajo blankets and silver jewelry to Pueblo pottery and baskets. Native American artists were also employed to decorate the hotels and stores of the Fred Harvey Company, together with craftsmen and -women practicing their art, in appropriate settings, in view of the tourists. The author of the *New York Times* article observed that the exhibition gave the impression that both sides benefited from the encounter, without any hint of the Indian being victimized in the exchange. In a speech given shortly after the exhibition opening by Rayna Green, director of the American Indian Program at the Smithsonian Institution, she said the Indians of the Southwest had already "learned to play Indian from the 17th century onward, first from the Spanish." The *Times* article closed with a chilling quotation from a video about Native Americans recalling the glory days of the Fred Harvey Company. What the company did, said a seventy-year-old Zuni, was to take them "from ritual to retail."

Strolling in the plaza, peeping into every shop and art gallery, what did I experience, precisely? Yes, the atmosphere of an invented romantic Spanish colonial past was perhaps too well maintained, and the artists (from early on, Santa Fe, and Taos, our next destination, were marketed as artist colonies) and Indians were there, playing the tourist game in a rather dignified and ironic way. It didn't disturb me particularly: after all, I was from Venice, a city that has been surviving mainly as a tourist attraction for centuries, selling its own atmosphere of glorious art, architecture, death, and decay. I was

Taos Pueblo (with friend of the author), New Mexico

used to sharing the narrow Venetian *calli* with masses of tourists unaware of the rules governing the navigation of the labyrinthine urban fabric; to watching vegetable stalls and bakeries disappear daily, giving way to souvenir shops; and to explaining patiently that no, Ponte Vecchio is in Florence—what you are looking at is the Rialto Bridge and no, I don't own a gondola.

Taos Pueblo: Freezing cold, dramatic sky, intense, fierce light, and clouds throwing unexpected shadows. Timeless profiles of buildings and mountains, wood fires perfuming the air with the aroma of piñon and sage. We were stopped at the entrance by a polite man: there was a fee for the use of each of our cameras, and we were told to ask permission to take photos of the inhabitants. Few people were around, most of them indoors, their attitude unaffected and remote, welcoming tourists in uncluttered adobe interiors transformed into shops. We were the only visitors that day. We wandered around without exchanging impressions, silent—almost speechless in fact. We didn't photograph the inhabitants of Taos, and when I go through the pictures taken during that visit, the only human figure against the stunning landscape is that of a solitary French historian.

I was aware of the many architects who had preceded us on such a pilgrimage, including Rudolph Schindler, who in 1915 confided to Richard Neutra: "My trips to San Francisco and among Indians and cowboys are unforgettable experiences. That part of America is a country one can be fond of, but the civilized part is horrible, starting with the President down to the street-sweeper."[10] Schindler considered Pueblo architecture the only true indigenous architecture he had seen in the United States: "The only buildings which testify to the deep feeling for the soil on which they stand."[11] Upon his return to Chicago, he proposed to design for a Dr. Martin in Taos a country house in adobe construction. Dr. Martin's house was never built, but the "lesson" of Pueblo architecture remained a considerable if subtle presence in the development of his California modernism. His friend Neutra had a similar attitude. Neutra saw adobe architecture for the first time reproduced in the Museum of Natural History in New York in 1923 and praised Pueblo Indians for being "the people who influenced the modern California building activity."[12]

Their feelings present an interesting contrast to the view one of their great American "masters," Frank Lloyd Wright, who dismissed "Indian or Mexican 'hut' builders." For all of Wright's love for the "organic" and his poetic vision of buildings as "shelter," in his opinion architecture, like music and literature, was beyond the Hopi. For him, the native way of building was not even sympathetic to the environment: "The Indian Hopi house is no desert house with its plain walls jumping out to your eyes from the desert forty miles or more away."[13] I was also thinking about Aldo van Eyck and his ethnographic investigation of the architecture of the Dogon of West Africa and of the Amerindians of New Mexico, which he visited in 1961, trying to remember if any of them had made remarks about photography.

What came to mind was a chapter tellingly titled "The Inscrutable," from Reyner Banham's *Scenes in America Deserta*. Like us, he came for the first time to Taos Pueblo in winter and found the place deserted, the central plaza empty. Like us, he concentrated his "photographic attention" on the "memorably strong elementary buildings" as "so many, many architectural visitors have done." And then, in an arresting passage, Banham explained how he found it impossible to take a picture:

> Trying to pursue surviving photographic light, I probed the terraces through the zoom lens until I suddenly came upon a scene that I could not bring myself to photograph. High on the terraces there was a white-robed figure, looking almost like a Roman statue, who appeared to be addressing the westering sun. I knew nothing about the priests of Taos at the time; his garb was unexpected and his action inscrutable. I felt, overwhelmingly and in a way that was new to me, that I had seen a piece, a small corner, of a culture that felt more alien, unknown than anything I had encountered before. The sense of having come up against a glass wall through which seeing was possible but comprehension was not […] has never really gone away ever since.[14]

At the time I didn't know precisely in what climate Banham wrote this extraordinary statement. *Scenes in America Deserta* was published in 1982, more then a decade after Banham's first encounter with the native inhabitants of the Southwest. I felt that his was the only acceptable stance, against a depressing panorama of more of a century of well-meaning travelers ready to embrace Indian culture and offer their own questionable and self-serving interpretations.

Taos: I knew about the town of Taos through the writings of the ailing "overcivilized" intellectuals and artists who had come there between the two world wars to seek solace and renewal in the purified, dry desert air and in the rituals performed by "primitives" still living at one with Nature and the Gods. The American heiress Mabel Ganson Evans Dodge Sterne arrived to seek "Change" with a capital "C," as she wrote in *Edge of Taos Desert,* the fourth and last volume of her autobiography. She came to join her third husband, the painter Maurice Sterne, who wrote her a prophetic letter in November 1917: "Dearest Girl, Do you want an object in life? Save the Indian, their art-culture—reveal it to the world!...That which Emilie Hapgood and others are doing for the Negros, you could, if you wanted to, do for the Indians, for you have the energy...and, and above all, there is somehow a strange relationship between yourself and the Indians."[15] And indeed she devoted her immense vigor and financial resources to save "her" Indians, spending the rest of her life at Taos, building, together with her new husband, the Pueblo Indian Antonio Luhan, a mythical adobe house designed to become "a kind of headquarters for the future... [and] a base of operations for really a new world plan."[16]

There, the new and "whole" Mabel Dodge Luhan managed to attract and enlist to her cause an astounding number of leading figures of the postwar American and European intelligentsia: painters Andrew Dasburg, Marsden Hartley, and Georgia O'Keeffe (who later set up her own house at Abiquiu); photographers Paul Strand, Ansel Adams, Edward Weston, and Laura Gilpin; stage designer Robert Edmond Jones; choreographer Martha Graham. A sojourn with the Luhans inspired Willa Cather to write the thoughtful and delicate *Death Comes for the Archbishop* (1927), and Mary Austin embarked on a trajectory that changed her life. A writer already familiar with the semi-arid country of south-central California and with the Paiute and Shoshone Indians, Mary Austin first arrived in Taos in 1919 and visited frequently, studying northern Pueblos and becoming involved in a famous controversy about the ownership of Indian lands. In 1924 she settled permanently in Santa Fe, helping to organize the Spanish Colonial Arts Society for the promotion and preservation of Hispanic artistic traditions and using her own home as an operations center for the foundation of a new America. The arid Southwest was to be the setting "for the next fructifying world culture" because its climate had shaped ideal "American" communities: egalitarian, environmentally conscious, producers of "adequate symbols in art," and still practicing meaningful religious rituals.

Progressive social reformer John Collier, another early visitor to the Mabel Dodge Luhan house in Taos, stayed to become the "greatest Indian Commissioner" in the history of the United States. He launched his crusade to defend the lands and rights of the Pueblos with an essay entitled "The Red Atlantis," joining an expanding circle that promoted a cultural nationalism rooted in regionalism. Anthropologist and folklorist Elsie

Clews Parsons, another friend of the Luhans, fought along the same lines to preserve Native American art, ritual, and social organization as an alternative to a deracinated, neurotic Anglo-Saxon civilization. She also took advantage of the friendship of the Indians to publish information about their cults that they wished to keep secret, following in the footsteps of many ethnographers before her.

D.H. Lawrence came to Taos, also lured by Mable Dodge Luhan, fleeing a Europe devastated by mechanized war, to establish his utopia and immerse himself in the "oceanic" feeling of the primitive. His was an ambiguous, uneasy, encounter: the "old red forefathers" were devoted to a "cult of water-hatred" and never washed "flesh or rags." Their drumming and dancing resonated in the deepest recesses of his overly sophisticated European soul, evoking an ancient communion with the gods and nature, but

Approach to Shiprock, New Mexico

signaling, at the same time, the impossibility of its recovery for the civilized man. At the conclusion of the depiction of his first experience with the Navajos' ritual dancing, Lawrence wrote: "I have a dark-faced, bronze-voiced father far back in the resinous ages. My mother was no virgin. She lay in her hour with this dusky-lipped tribe-father. And I have not forgotten him. But he, like many an old father with a changeling son, he would deny me. But I stand on the far edge of their firelight, and am neither denied nor accepted. My way is my own, old red father; I can't cluster at the drum anymore."[17]

This impossibility was explored in its most grotesque ramifications in *Brave New World,* the ominous science fiction novel written in 1932 by Aldous Huxley, before his own visit to Taos, on the basis of conversations in Italy with Lawrence. He depicted a future world ordered as castes of bottle-produced individuals, conditioned to like the work they are destined to perform, made happy by the government-controlled drug "soma" and practicing compulsory, orgiastic, meaningless sex. Only on a New Mexico reservation, surrounded by barbed-wire fences, a few thousand Indians are left to live a "savage life." Two tourists from the "civilized" world visit the reservation to observe with mounting disgust the filthiness and squalor of the Indians' existence. Puzzled and repulsed by the lack of hygiene, the sight of women actually giving birth, familial relations, and hideous ceremonies—Huxley here offers a fanciful portrayal of regional religious ceremonies, mixing Navajo rituals with the Hopi Snake Dance and the Spanish Penitentes' practice of self-flagellation—the tourists bring one of the "savages" to the civilized world as an object of curiosity. The novel ends with the suicide of the rescued savage, unable to fit into the technologically controlled, consumerist "happy" society that he finds inhuman and revolting.

I did find the paternalistic, well-meaning, but eventually exploitative and even racist attitudes of these early twentieth-century disillusioned (with Western culture) intellectuals much more disturbing than the straightforward commercialization of entrepre-

Goosenecks, Utah

neurs such as Fred Harvey. Nevertheless, they left quite a mark on the region, and their houses and the landscapes they described, painted, and photographed have become major tourist attractions. One can visit Mabel Dodge Luhan's house; Ghost Ranch, where D.H. Lawrence lived with his wife, Frieda; the chapel where his ashes are supposedly preserved; the residence of Georgia O'Keeffe in Abiquiu, and Brett House (home of the painter Dorothy Brett, the only member of Lawrence's utopian community) that at the time of our visit had become an upscale restaurant. Tourist brochures publicize the "stunning O'Keeffe country" or invite you to plan excursions to "D.H. Lawrence's haunts" in and around Taos. We did follow the ritual of retracing their footsteps, at least in part, and I remember visiting the Kit Carson Home and Museum, and the house and studio of one of the cofounders of the Taos Society of Artists, painter Ernest Blumenschein.

But what I remember most about Taos is the overwhelming New Age atmosphere. I learned that as early as the 1980s, the number of alternative healers proposing mental and physical therapies (about 100) matched the number of artists residing in the town. Most of the New Age healers took inspiration from Indian and Hispanic practices and subscribed to the legend that mystical restorative forces are at work in the area. A lot of them, of course, were Jungians. This was something I knew about, as many scholars concerned with the American Southwest refer to the heavy presence of Jungians in Taos.[18] Architectural historian Vincent Scully, for example, in his monumental *Pueblo: Mountain, Village, Dance* (1975), observes: "Taos attracts Jungians, especially, like flies to compost, and indeed everyone who is attracted to the mystery of humanity's buried thoughts."[19]

Carl Gustav Jung was one of the early visitors to Taos, a big catch of the infatigable Mabel Dodge Luhan, a Jungian herself. Jung went to sit at the feet of the priests of Taos Pueblo to gather a new perspective on the psyche of the white man, and more material to support his theory of the archetypes and of a collective unconscious. Despite the apparently disparaging remark, Scully himself seems to follow in the footsteps of

Monument Valley, Navajo Indian Reservation, Arizona/Utah border

Jung, proposing a parallel interpretation of Indian rituals and Greek tragedy. In the preface to his volume on the Pueblo, Scully presents the research, largely based on ethnographic literature, as the extension of his study for *The Earth, the Temple, and the Gods: Greek Sacred Architecture,* a book he published in 1962. The analysis of the Pueblos, writes Scully, "grew directly out of my previous work in Greece, whose landscape the American Southwest strongly recalls, not least in the forms of its sacred mountains and the reverence of its old inhabitants for them. Only in the pueblos, in that sense, could my Greek studies be completed, because their ancient rituals are still performed in them. The chorus of Dionysus still dances there."[20] Reyner Banham, in *Scenes in America Deserta,* describes the effort of Scully as "the most splendid and disastrous of all paleface attempts to focus on 'the Indian phenomenon'."[21] Scully's "flights of fancy," explains Banham, were to some extent acceptable in the case of Greece, where he went equipped as a scholar trained in a classical tradition greatly indebted to Greek civilization. With regard to the Pueblo and their culture, which Scully knew only in "translation," he was utterly missing the mark in attempting the comparison between "polis"

and "pueblo." In this controversy I found myself on the side of Banham, even if Scully provides at least an interpretation, while I was at loss—like Banham, fascinated but incapable of comprehension. And still I had not seen the Indians dancing.

Toward the conclusion of his extended critique of Scully, Banham oddly remarks: "What the book does deliver is photography (much of it his own) that has the unmistakable ring of truth."[22] Is photography always truthful, and does it explain anything? One would expect a subtler comment from as thoughtful and keen an observer as Banham. In fact, his statement is also inaccurate: Scully elucidates in the preface of his volume that he had to use a great number of old photographs because of restrictions already in place in numerous communities. Photography was forbidden in the Hopi and Keres towns. The Zuni villages Taos and Acoma permitted photographing of the towns, but never of the

dances. The prohibitions made his task very difficult, but Scully respected them: "We can only be glad that the surviving Americans became so canny at last. Otherwise, one is soon doing it for the camera rather that for the god, and that is the end of it all."[23]

The interdictions included (and often still include) sketching, filming, and taping, and Scully is not the first scholar to note them. The earliest ethnographic reports from the Southwest, including the famous (or infamous) narrative of Cushing, insist on the Indians' active, if hopeless, resistance against any form of representation of themselves and their ceremonies. In spite of this unwillingness, scientists, journalists, militaries, missionaries, tourists, and professional photographers systematically captured their physiognomies and most sacred rituals on camera. Some photographic reportages were conducted with the best intentions, even if with the utmost disregard of Indians' beliefs and feelings. Edward S. Curtis's epic project of documenting the "vanishing race" is a case in point.[24]

Equally momentous in the field of art history were the photographic records collected in New Mexico and Arizona by Aby Warburg at the end of the nineteenth century.

Oddly neither Banham nor Scully mentions the visit of the German scholar and the crucial role it assumed in the development of his "pathos formula" and the Dionysian impulse in the arts. Warburg went to the Southwest after a number of conversations with ethnographers at the Smithsonian in Washington. He registered in his journal the Indians' displeasure with photography, but went on taking and buying pictures. At the same time, he mourned the killing of the primordial vitality and unity expressed in Indian rituals, an irreparable loss brought about by the implacable scientific and technological character of the schizophrenic European "civilization."[25]

In a recent essay Beverly Singer, professor of anthropology and Native American studies at the University of New Mexico, refers to a renewal of interest in Indian photographic portraits in the 1970s, which led to a revival of collecting everything native.[26] The late 1960s and early 1970s were the years during which Scully and Banham con-

Clouds dissolving over the Grand Canyon after a winter storm, Arizona

ducted their explorations of the American Southwest. During this period, Banham explains, "Indian culture was to be admired as an exemplar to wasteful and ecologically destructive Western man."[27] It must have been precisely the time of the epic migration of hippies from the birthplaces of the counterculture to the American Southwest. Leaving Haight-Ashbury in San Francisco or Lower Manhattan, places overrun by junkies and other ugly characters and constantly exploited by the media, the flower children converged on the arid and exotic territories of Colorado, Arizona, and New Mexico in search of free (or cheap) lands on which to experiment with alternative, communal ways of life. New Mexico, and Taos in particular, became the epicenter of the phenomenon. In 1969, Stewart Brand, editor of the *Whole Earth Catalog,* forum of the dispersed tribes of the counterculture, proclaimed: "New Mexico is the center of momentum this year and maybe for the next several. More of the interesting intentional communities are there. More of the interesting outlaw designers are."[28]

The statement introduced the presentation of the mythical Alloy conference, an event that took place during the spring equinox of the same year in an area situated between the Mescalero Apache Reservation and the Trinity atomic bomb test site. According to Brand, the initiator of the conference was Steve Baer, inventor of the Zome, a variation on the geodesic dome of Buckminster Fuller, which became a favorite model of construction in newly founded countercultural communities. What Baer had in mind, explained Brand, was "a meld of information on Materials, Structure, Energy, Man, Magic, Evolution, and Consciousness."[29] Given this premise, the choice of the site for the conference was strategic, reflecting the interests not only of Baer (who moved to Albuquerque after studying mathematics at the ETH in Zurich) but of most of the participants. In fact, many of the 150 "outlaw designers" at the conference shared a fascination for the sciences and advanced technologies, including ones developed by and for the military, and a profound interest in Native American culture—and not just because of

the exemplary ecological attitude evoked by Banham. What attracted the generation of followers of LSD prophets and gurus expounding the wisdom of exotic religions were the "magic" of the Indian system of beliefs and the spiritual practices involving the consumption of drugs.

Typical is the case of Stewart Brand who, after studying ecology at Stanford, served in the U.S. Army, became involved in the work of USCO ("U.S." company), an anonymous group of East Coast artists producing avant-garde multimedia installations. Brand then moved to San Francisco to become a member of the Merry Pranksters, the crazy tribe of Ken Kesey, responsible for organizing the notorious Acid Tests. In the early 1960s, while collaborating with USCO, Brand visited Warm Springs, Blackfoot, Navajo, Hopi, Papago, and other Indian reservations to research and gather photographs and other materials for a multimedia experience called "America Needs Indian." The event employed movie projectors, Indian dancers, and multiple soundtracks playing simultaneously. In 1966, it became part of the Trips Festival in San Francisco, one of the greatest countercultural moments. Brand, who for a time was married to a Native American mathematician, mentions his encounters with the Native Americans in the *Whole Earth Catalog* as introduction to a recommended collection of publications written on Indians or by Indians: "The booklist that follows comes from two intense informal years (and five slack ones) hanging around Indian reservations, anthropologists, and libraries. Long may Indians, reservations, anthropologist sand libraries thrive! They gave me more reliable information, and human warmth, than dope and college put together. I am sure the books all by themselves cannot deliver The Native American Experience. For that you need time immersed in the land and neighborly acquaintance at least with some in fact Indian."[30] He was preaching to the converted. Members of the counterculture in the Southwest were already fraternizing with Native Americans, displaying an active interest in particular in peyote ceremonies, living in tepees, wearing Indian attire, and adopting names like New Buffalo for their new communities.

They were also rediscovering the previous generation of escapees and Indian lovers, from D.H. Lawrence to Aldous Huxley and Mabel Dodge Luhan. In the cult film *Easy Rider* (1969), the tragic account of a journey of two countercultural bikers traveling from Los Angeles to New Orleans in search of America (including a fictional portrayal of New Buffalo), one of the characters, played by Jack Nicholson, constantly quotes D.H. Lawrence. Actor Dennis Hopper, after the incredible success of the film, moved to Taos and lived in the house of Mabel Dodge Luhan with the hope of creating an alternative movie center.

This enthusiastic espousal of Indian costumes and way of life was inspired more by a fanciful image of the Native Americans than by the reality of local tribal traditions. The tepee, for example, was far from the typical housing of the region. The Navajo built hogans and the Pueblo, adobe architecture. Likewise, names such as New Buffalo evoked more the hunting and nomadic life of the tribes living on the plains than the sedentary habits of the Pueblo, who subsisted mainly on a diet of corn, beans, and squash. Scholarly books, diaries, memoirs, and oral narratives document these encounters and the tolerant attitude of the Native Americans.

In the eyes of many whites, an alliance was forged between hippies and Native Americans. Years later, Brand noted: "By the end of the '60s, Indians had been adopted by the hippies, and to everyone's astonishment, not least mine, it basically worked out. There was a transmission of traditional frames of reference from older Indians to hippies, who were passing it to their young peers in the reservations and a lineage was inadvertently, but I think genuinely, preserved."[31] But what was the Indian perception of this supposed alliance? And did it really take place? Scully, in *Pueblo: Mountain, Village, Dance,* offers a glimpse on the Indians' response, reporting an episode that took place in June 1968 at Shipaulovi. Hopi clowns were performing during the intervals of a katchina dance "satirizing social workers and the agents of the Bureau of Indian Affairs. At other times they have taken off hippies and missionaries, tourists, and especially all Indian lovers, always."[32]

The year after our trip, Philip Deloria, a historian of Indian descent, published *Playing Indian,* a thoughtful investigation of the way Americans, since the time of the

Boston Tea Party, have repeatedly appropriated Indian dress and acted out Indian roles. Time after time, Americans have been "playing Indian" to shape their national identity. Retracing this fascinating history, Deloria devotes an entire chapter to the counterculture Indians and New Age adherents, describing also the response of real Indians. In conclusion, Deloria observes: "Like many before them, they had turned to Indianness as sign of all that was authentic and aboriginal, everything that could be true about America. [...] Yet like those who came before, they found that Indianness inevitably required real native people, and that those people called everything into question."[33] However, despite all of the misunderstandings, inconsistencies, and paradoxes of the encounters between hippies and Indians, the years of the counterculture—of the revolt against the dominant values of American society, and of the civil rights battles—had a profound impact on Indian consciousness. As I was to learn later, in the unrest of the time American Indians found the seeds of a transformation that has been compared to a cultural revolution. But in the winter of 1997, during the journey that took us to the Zuni and Hopi towns high on the mesas, to the bare and silent remains of Canyon de Chelly, Mesa Verde, and Chaco Canyon, to the ominous museums of Los Alamos, to the unimaginable gorgeousness of the Grand Canyon, Monument Valley, and Shiprock, and to the reservations sprinkled with casinos and dialysis clinics, the fruits of that revolution were still unseen. The Indians, selling souvenirs, acting as guides, and living in evident poverty, remained a baffling presence. And then, at the end of our trip, we saw them dancing.

Acoma: The "Sky City," almost an afterthought. One of us insisted on visiting it even if it was our final day and we had to catch planes going in different directions early in the afternoon. We left the last of the Best Western hotels early in the morning. It was still dark and exceedingly cold. We had to leave our car at the foot of the mesa where Acoma has stood, unchanged, for centuries. A guide drove us up in the astonishing radiance of the morning. Elemental adobe compositions, blinding sunshine on snow and ice, terse and freezing sky, drums and stamping feet—it was December, time to celebrate the winter solstice. Once more, we were the only tourists. We sat, unused cameras in our hands, in a corner of the church San Estevan del Rey. Dressed in traditional attire and beautifully masked, the men came, and the adolescent boys, and the young women, and the mature women and the children, joyously dancing, honoring the bountiful new year to come. Again we were speechless, a silence that stayed with us beyond the quick adieus at the airport. For the first time in my life, I had felt the unbelievable power of a traditional society, and the experience still haunts me more than ten years later.

CODA: 2007

In the early 1950s, one widely advertised attraction of Las Vegas was its proximity to the Nevada Test Site. An iconic 1957 photograph of "Miss Atomic Bomb," portraying showgirl Lee Merlin of the Sands Hotel, with a cotton mushroom cloud added to the front of her swimsuit, embodies the spirit of the time. One can still buy souvenirs displaying the blond Merlin, arms outstretched, euphorically celebrating the Atomic Age. Las

Vegas, the city of "sin," was strangely gaining a new legitimacy by joining the Cold War effort and transforming the specter of nuclear annihilation into spectacle. Documents about the Las Vegas of the time, such as the famous postcard advertising the Pioneer Club (circa 1955) with its winking cowboy sign and a glowing red mushroom cloud in the distance, show how images related to the burgeoning atomic tourism were quite often associated with a preexisting strategy of capitalizing on the pioneering era and Native American past of the area. After reducing to entertainment the painful history of conquest and domination over the western territories and their indigenous occupants, Las Vegas was performing the same operation on the Cold War and the threat of obliteration of life and civilization: the tragedies and perils of the old and new wars reassuringly contained and gloriously reframed by the powerful American myth of the frontier.

In 2005 the Atomic Testing Museum opened in Las Vegas. An affiliate of the Smithsonian Institution, the museum is located only a mile from the Las Vegas Strip and appears to be a popular tourist destination. The mission of the museum is to present scientific matters in a compelling way, preserve the legacy of the Nevada Test Site, and promote public accessibility and understanding of the site. Various galleries document the history of the site in the context of the Cold War, display how the Atomic Age was reflected in pop culture, show photographs, films, and interviews with workers and protestors. Not far from the theater (a replica of a bunker) are situated the Steward of the Land Galleries I and II. The first covers geology, hydrology, and radiation monitoring. The second is dedicated to archeology, endangered species, and Native Americans. According to the museum authority, the collection illustrating crafts and other objects used by ancient inhabitants is being completed with the collaboration of the local tribe. Nuclear power and American Indians: at the Atomic Testing Museum, carefully reframed and updated, we find the association already explored and exploited by the Las Vegas of the 1950s. At the museum, the Indians, instead of being presented like the warriors of Buffalo Bill's show, are offered to the visitors as the descendants of a primeval civilization living in harmony with the arid territory. The label "stewards of the land" seems to suggest the possibility of the reclamation of a technologically devastated terrain thanks to the wisdom of the original inhabitants. A similar strategy is deployed at the Nuclear Test Site itself, which has also become a tourist destination. Signs posted on the fence surrounding the NTS, after describing the origin and function of the area, tactfully announce: "Archeological studies of the NTS area have revealed continuous occupation by prehistoric man from about 9,500 years ago. Several prehistoric cultures are represented. The last aboriginal group to occupy the site was the Southern Paiute who foraged plant foods in season and occupied the area until the arrival of the pioneers."

Once again Americans are playing Indian, or more aptly, playing with the Indians. The Native Americans represented in the museum and mentioned on the NTS signs are not the contemporary inhabitants of the reservations living in poverty next to contaminated areas, suffering from obesity, diabetes, heart disease, and alcoholism, making an uncertain life catering to tourists. The lands taken from the original owners are symbolically "returned" by the institutions, but not to the Indians of the present, transformed by the reality of contemporary America. The reinstated Indians offered to the gaze of the tourists are safely frozen in time. They are the custodians of immemorial knowledge, the captives of tradition and authenticity.

Tradition and authenticity are the traps that a new generation of Native American artists are exposing and trying to evade. They challenge the carefully constructed prison where they are condemned to conform to a required stereotype. Their weapon of choice is often photography. From among many provocative artists, I will mention only three examples. In 2005, Zig Jackson became the first Native American photographer represented in the collection of the Library of Congress in Washington, D.C. Jackson donated four prints from each of three series. The first group, under the title "Indian Photographing Tourist Photographing Indian," represents, humorously, invasive tourists taking pictures of reservation Indians. The second, "Native American Veterans," more somberly honors military veterans and their families from Plains Indian reservations. "Entering Zig's Indian Reservation," the last, darkly amusing series, represents Jackson himself. Wearing Indian attire and sunglasses, he poses at various sites in San Francisco next to a huge, official-looking sign that says: "Entering Zig's Reservation." Under the heading, the sign lists private property rules that include "No Picture Taking," "No Hunting," "No Air Traffic," and "New Agers Prohibited."

Grand Canyon after a winter storm, Arizona

NOTES

1 Hulleah Tsinhnahjinnie, a Diné/Seminole/Muscogee, an artist who privileges photography as favorite medium and conduit of political expression, became internationally famous with "The Damn Series" of 1997. When exhibited at the Barbican Gallery in London, two images in particular captured the attention of the audience and the press: *This is not a commercial, this is my homeland,* and *Damn! There goes the Neighbourhood.* The first depicts Monument Valley, the iconic Southwestern panorama of mesas employed innumerable times as a setting in advertising and film. The superimposed inscription removes it from the realm of clichés and reframes the iconic scenery as sovereign Diné land. The second represents a desert with, in the foreground, an Indian warrior portrayed in an old photograph with a smoking gun in hand, and a garish, bullet-ridden Oscar Mayer Wienermobile in the background. Once again, the inscription that seems to come, cartoonlike, from the mouth of the warrior eloquently decries the fate of the Indian people and the lands they have lost.

In 1992, James Luna, a Luiseño Indian, proposed at the Whitney Museum in New York a performance titled *Take a Picture with a Real Indian.* Visitors were asked to pick a real Indian from a selection of cardboard cutouts and take a Polaroid. The work was inspired by a trip through Navajo land during which Luna had seen Indians selling souvenirs and catering to the tourists. A few years before, Luna, with *Artifact Piece,* had spectacularly called attention to the exhibition of Native American people and their relics by displaying himself in a glass case at the Museum of Man in San Diego. For days he remained motionless, dressed in a loincloth and surrounded by personal documents and ceremonial objects. Many members of the public were stunned by the discovery that the unmoving figure on exhibit was actually a living and breathing individual. In another memorable performance, *Petroglyphs in Motion,* Luna presented a nonlinear history of Native American man. Beginning with a petroglyph, Luna in turn impersonated Shaman, Rockabilly, War Veteran, Drunk, and Coyote. Vertiginously traveling through time, the characters mutate, learn, and evolve.

The works of these artists are powerful and self-explanatory. They speak eloquently of a new form of resistance and self-representation. The camera, held for so long in the hands of the white man, the scientist, the missionary, the military, and the tourist, is no longer kept at bay, albeit with interdictions often ignored. Photography, now in the hands of American Indians, is no longer there to record stereotypes, immortalize tradition, and confirm authenticity. Poignantly or ironically, it exposes unbalanced systems of relationships, different perceptions of time, history, and reality. The Indian wars have moved to new battlefields. To paraphrase James Luna, who in 2005 (with Ed Ruscha) represented the United States at the Venice Biennale: tourists beware—the petroglyphs are in motion.

Benedetta Cestelli Guidi and Nicholas Mann, eds., "Excerpts from Aby Warburg's Diary," in *Photographs at the Frontier: Aby Warburg in America, 1895–1896* (London: Merrell Holberton Publishers with the Warburg Institute, 1998), 155.

2 Susan Sontag, *On Photography* (New York: Picador, Farrar, Straus and Giroux, 1977), 23–24.

3 Paul Chaat Smith, "Luna Remembers," in Truman T. Lowe, Paul Chaat Smith, curators, *James Luna: Emendatio,* 51st International Art Exhibition, La Biennale di Venezia (Washington, D.C.: Fondazione Querini Stampalia, Smithsonian Museum of the American Indian, 2005), 26.

4 On nineteenth-century ethnography in North America, see Curtis M. Hinsley, *The Smithsonian and the American Indian: Making a Moral Anthropology in Victorian America* (Washington, D.C., and London: Smithsonian Institution Press, 1981); Curtis Hinsley, "Ethnographic Charisma and Scientific Routine: Cushing and Fewkes in the American Southwest, 1879–1893," in George W. Stocking, ed., *Observers Observed: Essays on Ethnographic Fieldwork, History of Anthropology,* vol. 1 (Madison: University of Wisconsin Press, 1983), 53–69.

5 For a contextual analysis of the interpretation of Ruth Benedict, see George W. Stocking Jr., "The Ethnographic Sensibility of the 1920s and the Dualism of the Anthropological Tradition," in George W. Stocking Jr., ed., *Romantic Motives: Essays on Anthropological Sensibility, History of Anthropology,* vol. 6 (Madison: University of Wisconsin Press, 1989), 208–276.

6 Paul Rabinow, *Reflections on Fieldwork in Morocco* (Berkeley/Los Angeles: University of California Press, 1977); Edward Said, *Orientalism* (New York: Pantheon, 1979); James Clifford, *The Predicament of Culture: Twentieth-Century Ethnography, Literature, and Art* (Cambridge, MA: Harvard University Press, 1988); James Clifford and George Marcus, *Writing Culture: The Poetics and Politics of Ethnography* (Berkeley: University of California Press, 1986); George E. Marcus and Michael M.J Fisher, *Anthropology as Cultural Critique* (Chicago/London: University of Chicago Press, 1986); Roy Frank Ellen, ed., *Ethnographic Research: A Guide to General Conduct* (London: Academic Press, 1984).

7 Among the many studies see, for example, John Urry, *The Tourist Gaze: Leisure and Travel in Contemporary Societies* (London: Sage Publications, 1990); Tom Selwyn, ed., *The Tourist Image: Myth and Myth Making in Tourism* (Chichester/New York: John Wiley & Sons, 1996); Ruth B. Phillips and Christopher B. Steiner, eds., *Unpacking Culture: Art and Commodity in Colonial and Postcolonial Worlds* (Berkeley/Los Angeles/London: University of California Press, 1999).

8 Chris Wilson, *The Myth of Santa Fe: Creating a Modern Regional Tradition* (Albuquerque: University of New Mexico Press, 1997).

9 Katheleen L. Howard and Diana F. Pardue, eds., *Inventing the Southwest: The Fred Harvey Company and Native American Art* (Flagstaff, AZ: Northland Publishing, 1996). See also Martha Weigle and Barbara A. Babcock, eds., *The Great Southwest of the Fred Harvey Company and the Santa Fe Railway* (Phoenix: Heard Museum, 1996); Sandra D'Emilio and Suzan Campbell, eds., *Visions and Visionaries: The Art and Artists of the Santa Fe Railway* (Salt Lake City: Peregrine Smith Books, 1991); Virginia L. Grattan, *Mary Colter: Builder upon the Red Earth* (Grand Canyon: Grand Canyon Historical Association, 1992).

10 Schindler letter to Neutra, February 9, 1915, quoted in Thomas S. Hines, *Richard Neutra and the Search for Modern Architecture: A Biography and History* (Los Angeles/Berkeley: University of California Press, 1994), 321, note 310.

11 Rudolph Schindler, quoted in Elizabeth A.T. Smith, "R.M. Schindler: An Architecture of Invention and Intuition," in Elizabeth A.T. Smith and Michael Darling, eds., *The Architecture of R.M. Schindler* (Los Angeles: Museum of Contemporary Art, 2001), 20.

12 Neutra to Dione Neutra, n.d. 1923, quoted in Hines, *Richard Neutra and the Search for Modern Architecture*, 46.

13 Frank Lloyd Wright, "Organic Architecture," *Architects Journal* (August 1934), quoted in Paul Oliver, ed., *Shelter and Society* (New York: Frederick A. Praeger Publishers, 1969), 16.

14 Reyner Banham, *Scenes in America Deserta* (first edition 1982) (Cambridge, MA: MIT Press, 1989), 119–120.

15 Quoted in Mabel Dodge Luhan, *Edge of the Taos Desert: An Escape to Reality* (Albuquerque: University of New Mexico Press, 1987), xi–xii.

16 Quoted in Lois Palken Rudnik, *Utopian Vistas: The Mabel Dodge Luhan House and the American Counterculture* (Albuquerque: University of New Mexico Press, 1996), 93.

17 D.H. Lawrence, "Indians and an Englishman," in Keith Sagar, ed., *D.H. Lawrence and New Mexico* (Paris/London: Alyscamp Press, 1995), 10.

18 On Jung, Jungians, Indians, and New Agers, see Marianna Torgovnick, *Primitive Passions: Men, Women, and the Quest for Ecstasy* (New York: Alfred A. Knopf, 1997); Richard Noll, *The Jung Cult: Origins of a Charismatic Movement* (Princeton: Princeton University Press, 1994).

19 Scully, *Pueblo: Mountain, Village, Dance* (Chicago: University of Chicago Press, 1989), 84.

20 Ibid., xiii–xiv.

21 Banham, *Scenes in America Deserta*, 126.

22 Ibid., 127.

23 Scully, *Pueblo*, xvi.

24 Among the vast literature on Curtis, see, for example, Christopher Cardozo, ed., *Sacred Legacy: Edward S. Curtis and the North American Indian* (New York/London: Simon & Schuster, 2000). On Indians and photography, see also Martha A. Sandweiss, *Print the Legend: Photography and the American West* (New Haven-London: Yale University Press, 2002), chapter 6, " 'Momentos of the Race': Photography and the American Indian," 207–273; James C. Faris, *Navajo and Photography: A Critical History of Representation of an American People* (Salt Lake City: University of Utah Press, 2002).

25 Guidi and Mann, eds., *Photographs at the Frontier: Aby Warburg in America*. See also: Joseph L. Koerner, *Aby Warburg. Le ritual du Serpent: Art et Anthropologie* (Paris: Macula, 2003); Philippe-Alain Michaud, *Aby Warburg et l'image en movement* (Paris: Macula, 1998).

26 Beverly R. Singer, "Introduction," in Victor Masayesva Jr., *Husk of Time: The Photographs of Victor Masayesva* (Tucson: University of Arizona Press, 2006), ix–xviii.

27 Banham, *Scenes in America Deserta*, 120. Banham's first encounter with the American desert took place in February 1968.

28 Steward Brand, "Report from Alloy," *The Last Whole Earth Catalog* (New York: Portola Institute and Random House, 1971), 112.

29 Ibid.

30 Ibid., 382.

31 Quoted in Andrew G. Kirk, *The Whole Earth Catalog and American Environmentalism* (Lawrence: University Press of Kansas, 2007), 39.

32 Scully, *Pueblo*, 320.

33 Philip J. Deloria, *Playing Indian* (New Haven/London: Yale University Press, 1998), 180. An expanded and revised version of this chapter has been published as Philip J. Deloria, "Counterculture Indians and the New Age," in Peter Braunstein and Michael Doyle, eds., *Imagine Nation: The Counterculture of the 1960s and 70s* (New York/London: Routledge, 2002), 159–188. On Indianness and American society, see also Torgovnick, *Primitive Passions*.

ROBERT SUMRELL
AND KAZYS VARNELIS

7

QUARTZSITE, ARIZONA:

TALES OF DESERT NOMADS

One of the least known facts about the Mexican-American War of 1846–48 is that it gave rise to the short-lived experiment of the U.S. Camel Corps. That conflict made clear to military leaders that securing the rough terrain of the Southwest against Native Americans or the Mexican government, both unhappy with the growing power of the United States in that region, would not be easy. The majority of American casualties in the war fell victim not to enemy fire but to the harsh conditions, proving how inhospitable that region is to traditional cavalry and infantry.

Convinced tht they had found a solution, Second Lieutenant George H. Crossman, a veteran of the Seminole Wars, and Major Henry C. Wayne, a quartermaster, gained the support of Secretary of War Jefferson Davis and persuaded Congress to allocate $30,000 to the Camel Military Corps in 1855. Like many such ideas, this seemed sensible at the time. Arabia and the American Southwest are similar climatologically. Dromedaries are obviously well suited to the arid environment, and although it is hardly plausible that Crossman, Wayne, or Davis would have known this, giant prehistoric camels once roamed the continent. More likely, Crossman and Davis had heard of plans to bring camels to the Mojave desert as pack animals and might even have received word that the animals were being brought to Australia by settlers to help colonize the Outback. In what would be the first operational test of material in the field by the Army and inaugurate the tradition of military research, the Corps was charged with determining the capabilities of the animals for the possible formation of a camel cavalry, for deployment in artillery units, and for use as pack animals.

Two years later, the Army imported seventy-seven North African camels and a Syrian camel driver named Hadji Ali to the Southwest. Based at Camp Verde, the Corps was charged with establishing mail and supply routes to California to the West and Texas to the East. Although the camels thrived in conditions that would fell any horse, the experiment was not without its problems. The animals did not adapt well to the rocky terrain. They scared other pack animals such as horses and burros. Soldiers found them foul smelling and bad tempered and complained about camels spitting at them. Nevertheless, the new Secretary of War, John Floyd, was impressed and asked Congress for a further 1,000 camels. But tensions between the North and South were rising and the Congress couldn't be bothered with the distant lands of the Southwest. Moreover, upon being appointed Commander of the Texas Army, Major General David E. Twiggs, sometimes known as "The Horse" (but also as "Old Davy" or "the Bengal Tiger") was horrified to discover the Camel Corps in his charge and successfully lobbied Congress to be rid of the beasts. Perhaps it is just as well: Twiggs would soon surrender his command and, with it, the Texas Army to the Confederacy.

Instead of serving the Confederacy, in 1863 the Camel Corps was sold off at auction. Most of the animals would wind up in private hands, but some would be released into the desert, where they became feral. Hadji Ali, now known as "Hi Jolly," remained, although it is unclear whether this was to pursue the American dream or simply because he was marooned far from home. After a time running a camel-borne freight business, Hi Jolly married a Tucson woman and moved to the west Arizona town of Tyson's Wells, 9 miles west of Quartzsite, Arizona, where he worked as a miner until he died in 1902. In memory of his service, the government of Arizona built a small pyramid topped by a metal camel on his gravesite in the 1930s. Feral camels would be seen roaming the desert until the early 1900s.

More recently, a new monument to the camel sprung up about a mile away, this time made of automobile rims and mufflers, to announce the fulfillment of the 150-year-old vision of self-sufficient desert nomads roaming the West. During the scorching summer, Quartzsite is a sleepy town of 3,397 inhabitants, but every year between October and March, a new breed of nomad comes to descend upon the town as hundreds of thousands of campers bring their recreational vehicles (RVs) to Quartzsite. These "snowbirds," generally retirees from colder climates, settle in one of the more than seventy RV parks in the area or in the outlying desert administered by the Bureau of Land Management (BLM). The BLM and local law enforcement agencies estimate that a total of 1.5 million people spend time in Quartzsite between October and March, a mass migration that temporarily forms one of the fifteen largest cities in the United States. If all of these residents inhabited Quartzsite at once, the result would be a more populous urbanized area than Dallas, San Jose, or San Diego, and possibly even bigger than Phoenix or Philadelphia, America's fifth-largest city.

For a half century after Hi Jolly's death, the population of Quartzsite remained small, with only about fifty people living in the outpost town on a permanent basis. By the 1950s, however, snowbirds began spending the relatively mild winter months in the

area, and by the 1960s the seasonal population would swell to 1,500. Many of these winter travelers returned year after year, and some settled permanently. As the community slowly grew, businessmen and civic boosters formed the Quartzsite Improvement Association and created a gem and mineral show to encourage more winter travelers to come.

Today, Quartzsite makes a radical break with the surrounding emptiness. Although it rejects vertical density and permanence, Quartzsite proposes a new kind of super-dense sprawl, achieving a remarkable horizontal density as RV is parked next to RV. That they came at all is the result of the invention of this modern replacement for Hi Jolly's camels, the RV. In retrospect, the development of such a self-sufficient beast, capable of hauling a family and enough food and water to sustain it over long distances, seems almost inevitable.

In his 1896 essay "The Frontier in American History," Frederick Jackson Turner, the founder of American Studies, observed that the U.S. Census bureau considered the frontier closed in the 1880s. For Jackson, this could only result in an epochal shift in the American psyche. Until then, he argued, Americans could renew themselves in the primitive conditions of the frontier. The loss of a true wilderness experience meant that the individual no longer had a ready place for social regeneration.

The end of social regeneration on the frontier, however, paved the way for a new idea: recreation in exurbia. After Henry Ford built the Model T, his "car for the great multitude," large numbers of individuals would flee the city on a regular basis in search of the newly domesticated "nature." Ford himself believed that the Model T's principal use would be to enable families to enjoy the blessing of hours of pleasure in God's great open space. Auto camping grew rapidly after World War I. By 1922, the *New York Times* estimated that of 10.8 million cars, 5 million were in use for camping. Soon "auto-tents" designed to fit the Model T would be available, with trailers for the Model T to tow to follow.

At first, campers would simply park in empty fields or by the side of the road, but this led to confrontations with angry rural townsfolk, seeing their way of life threatened by these nascent exurbanites who, they feared, would one day colonize the countryside. Soon campgrounds or "trailer parks" sprang up to provide places to stay with other campers on the road. Although campers sought nature and escape from a fixed community, they also enjoyed sharing this experience. Unlike the metropolis, trailer parks were places of relative homogeneity—campers were generally middle-class WASPs—so campers were able to tolerate living in close quarters.

During the Depression and World War II, however, camper trailers acquired a stigma for their use as temporary shelters while their inhabitants worked transient jobs. With the return of affluence in the 1950s, however, Americans once again desired to travel the country in self-sufficient "land yachts," untethered by hotels, inns, or motor courts. By this point, the old campers were unsuitable. Not only did they have the unpleasant connotation of transient housing to overcome, but as the size of homes grew in suburbia, automobile-drawn campers seemed small and cramped. The solution was to integrate the automobile and the trailer, creating the continuous unit now known as the "motorhome" or "recreational vehicle." This new kind of vehicle was gen-

erally much larger than the campers of old and permitted other activities to take place while the unit was being driven. Moreover, in doing away with the automobile or truck hauling the camper, the RV is clearly a vehicle that cannot be employed for traditional forms of work. You cannot drive your RV to a workplace: it exists purely for a lifestyle of leisure and consumption.

RVs have continued to rise in popularity. Today one in ten American vehicle-owning households owns one. But the majority of RV owners are in their sixties and seventies now, and spent their formative years in the 1950s, coming of age in the decade that saw the greatest migration in American history as they moved their young families from city to suburbia.

At Quartzsite, snowbirds annually reenact the process of settling the suburbs, choosing a vacant spot to inhabit next to others just like themselves, thereby recapturing the treasured anonymity and sameness of that era. Since everyone is in a camper, everyone is equal. Pasts are unimportant and incomes matter little. As in the postwar suburb, architecture is of no importance at Quartzsite. There are different models of RV and even some fundamental differences in RV typology—the full-fledged land yacht, the persistent trailer, the converted van, the converted bus—and some units may cost $500,000 while others cost $5,000. Nevertheless, an RV is an RV, a premanufactured unit that is not that dissimilar from other units of its kind. Like a historic preservation district in a contemporary city, Quartzsite is composed of similar units and individual expression is kept to a minimum. Add a flag, some plastic chairs, even a mat of green Astroturf, but your RV is still just like everyone else's and you are, quite likely, five or ten feet away from your neighbor.

Although the RV might appear to be an ultimate manifestation of American individualism, like the trailer campers of the 1910s, RV'ers generally see themselves as part of a community. After all, Quartzsite is the largest gathering of their kind in the nation, assembled purely by the desire to collect together. But this is still a particularly contemporary idea of community. The seventy-odd campsites in Quartzsite are generally privately owned and charge a moderate daily fee for usage. In the private campsite, the RV'er does not participate in any governance, choosing to let the "gated community" of the campground host do the governing. A sizeable percentage of travelers opt out of these areas, "boondocking" on BLM land where it is possible to stay for free for up to two weeks. Campers frequently form small communities on BLM land on the basis of RV brand, extended family ties, or group membership. Individuals skilled in striking pieces of flint with other pieces of flint to make primitive tools and ornaments, HAM radio buffs, full-timers who have sold their homes and live only on the road, nudists, and the Rainbow children, attracted to the freedom of Quartzsite as they wander the country recreating the hippie lifestyle of the early 1970s, all seek the company of others like themselves. In this, they reproduce the clustered demographics of posturban America where groups of remarkably specific inhabitants are congealing into discrete communities.

But beyond these clusters—and also like the world that Quartzsite is a microcosm of—community at Quartzsite is based on trade. Stimulated by the model of the Quartzsite Improvement Association, nine major gem and mineral shows and more

than fifteen general swap meeting shows attract RV'ers to the area. Much like the young hipsters who populate the fashionable districts of cities such as New York, San Francisco, and Los Angeles, campers at Quartzsite generally don't work except as full-time consumers. Seemingly incongruous juxtapositions of ever more bizarre goods appear throughout the markets: fresh shrimp cocktails hundreds of miles from the ocean are available next to cow skulls. African sculpture is popular, a demonstration of Quartzsite's role in a global network of nomadic trade. After wandering for a time at a Quartzsite show like the "Main Event" or the "Tyson Wells Sell-A-Rama," one is gripped by the thought that even the merchants don't come to Quartzsite to make a buck. As more than one sign advertising a merchant's need for a wife makes clear, merchants are more interested in interacting with people than in making a buck.

It's no surprise then that the exchange of rocks remains central to the market economy at Quartzsite, a focal point of the markets and a major draw for visitors. Often obtained from the surrounding mountains during leisurely hikes, with minimal labor applied to their retrieval and processing, Quartzsite's rocks circumvent any notion of labor or scarcity. Nor are these rocks useful. At Quartzsite, the markets teach us of a new nomadic way of life beyond any idea of affluence or material desire. Instead, the products sold at Quartzsite's markets are bought and sold to facilitate social relation-ships, not because they are needed or desired for their own sakes.

Karl Marx wrote that the social character of a producer's labor is only expressed through the exchange of commodities. But if there is no labor to speak of involved in bringing these valueless rocks to sale, exchanging them is a way for Quartzsite's winter visitors to remind each other that they have escaped the capitalist system into a world in which they are nothing, make nothing, and do not need to labor. At Quartzsite the subject finally disappears into the system of objects, but instead of being a source of oppression, as they were in Marx's day, objects become a source of liberation as they once were in the Potlatch or under the never realized utopian vision of Communism: "From each according to his abilities, to each according to his needs!"

REFERENCE

Wallis, Allan D. *Wheel Estate: The Rise and Decline of Mobile Homes.* Baltimore: Johns Hopkins University Press, 1991.

VINCENT BATTESTI

8

THE SAHARAN OASIS PUT TO
THE TEST OF ITS LANDSCAPE:
THE JERID

Oases of the Sahara can be seen as landscapes, but can we reduce the oasian landscapes to simply what can be apprehended visually? I will introduce the notion of socioecological landscape to support the double dimension of these oasian landscapes as ecological and social products and constructions. Such an approach can be applied to the notion of landscape in general, not only to deserts.

For millennia, Saharan oases have been the centers of human presence in the biggest desert areas, which are a priori hostile places for people. These oases are built environments. As they have an anthropic origin, so they are living landscapes, shaped by local inhabitants, but also increasingly by tourism. The case of the Jerid region (in the southwest of Tunisia) offers a valuable example of these built environnments, and that of the Siwa oasis (in the northwest of Egypt) provides an interesting counterpoint.[1]

DEFINING OASES' ECOSYSTEMS: ARTIFICIALITY

Oasis landscapes excite the imagination and evoke multiple images as soon as we think about them: colonial-era illustrations (postcards, novel covers, colonial exhibitions' catalogs), comic strips (often featuring desert extremes), and scenes from movies (such as *Lawrence of Arabia*, by David Lean in 1962).

We can most simply define an "oasis" as a fertile spot in an arid environment. I suggest calling an "oasis" a combination of both settlement and cultivation, and a "palm grove" the specific cultivated zone of an oasis. This short definition, however, runs counter to the common Western sense of an oasis as a quiet, Edenic places of

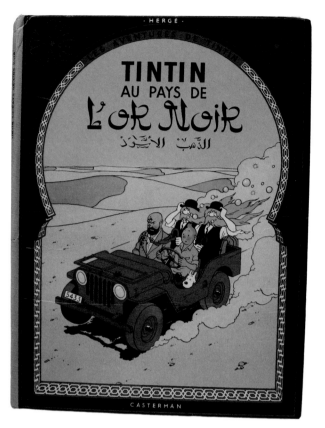

Tintin au pays de l'or noir (Land of Black Gold, Hergé 1950)

relaxation, where you can gather fruit from trees by stretching out your hand or kicking the trunk of a date palm (see the comic strip *Land of Black Gold,* by Hergé in 1950). Oasian landscapes suggest an easy lifestyle, perhaps because the oasis itself is a kind of a miracle in opposition to desert landscapes.

FACTORS NEEDED TO CREATE AN OASIAN LANDSCAPE

The miracle of the oases' ecosystem reflects their total artificiality. There is no oasis without human labor: an oasis is an exception in the desert, a kind of fragile bubble offering a specific landscape. And before everything else, it is agriculture. To exist, this agriculture requires the presence of three factors: water, plants, and labor.

It is rare that water rises by itself to the surface in the desert. For oases to bloom, it is generally necessary that people have the knowledge and means to use elaborate hydraulic equipment. Varied techniques have been implemented in the deserts of North Africa and the Middle East: wells— *foggara,* or *qanat* respectively, in North Africa and Iran—a system of kilometers of underground channels that collect drippings of water, and small dams. Sometimes water collected from distant mountains flows in rivers (*wadi*)—in south Morocco, for instance. The mere presence of water, however, does not guarantee the existence of oases. That is the case for the Kalahari Desert, where there is no big oasis despite the many streams from neighboring countries flowing into that vast basin.

The second factor is the presence of cultivated plants. In an oasis, plants need to be imported. The subsistence of local communities depends on cultivated cereals, vegetables, and fruit trees, all organized in mixed gardens. These plants have been domesticated from plant populations of nondesert regions that were originally adapted to different climatic and cultural situations. Even the emblematic date-palm tree, *Phœnix dactylifera*, was probably introduced in oases.

The third factor is human labor. Palm groves are a puzzle of small overlaid gardens. When oasian gardeners of the Nefzawa region (south Tunisia) prize the beauty of the neighboring Jerid oases, they refer immediately to the local agricultural skills and expertise: the oasis is understood as the outcome of know-how. Agricultural practices may vary from place to place, but permanent features remain, defining the oasis and contributing to the drawing of its landscape. For example, gardeners usually have to dig their plot of land by turning it upside-down with a hoe. In Jerid, a quarter of the garden is worked every year, which means that the entire surface of a palm grove is dug to a 50–70 centimeters' depth every four years. The essential irrigation also requires hours of labor and proficiency.

Work on the date-palm tree alone requires the careful planting of a female palm tree and its irrigation for years; the climbing of the trunk three or four times during the spring to open each inflorescence, and then pollination; the climbing of the trunk again during the summer to clean old palms, and to hang down the fragile date clusters so that they do not break; and finally the climbing of the trunk several times during fall to harvest the dates. We have to bear in mind that there are at least 200 date palms in a small half-hectare garden, in addition to other cultivation. The average garden plot is about half a hectare in the important Jerid palm groves (which total around 1,000 hectares). The oases of the Jerid region host the ideal type of palm grove, in the form of thousands of overlaid gardens, structured vertically. In most cases, a three-level structure is adopted: in the shadow of the palm trees grow fruit trees, which in turn cover vegetable or cereal cultivation. This structure creates its own microclimate, and along with the major components of the oases (irrigation system, etc.), it responds to the pedoclimatic conditions. These approaches were selected and implemented by generations of gardeners. The result of centuries of daily practices can be seen and analyzed today as the oasian landscape.

So, is the palm grove still a place of relaxation and ease? a site of visual enchantment for Romantic contemplation? I find strange the idea of tourism agencies inviting tourists to spend their holidays in what is in essence a kind of "plant factory." This mixed farming and intensive irrigated agriculture shapes the landscape; the spatial boundary of the cultivated zone is directly linked to the concern with maximizing water effectiveness and garden output.

Oasian landscapes are "built nature." But the question remains: by whom were they built? Are they only the construction of these generations of gardeners? We can reformulate the question "What is a landscape?" into "Who is talking about landscape, how, and why?"

A social anthropologist studies "local practices," but we should not limit our analysis to indigenous practices but broaden it to include all effective actors in these spaces. This means, in the Tunisian Jerid, actors of tourism and development. Those actors, with local people, converge to define what is an oasis, what "it is for," and how to read these spaces. This polyphony is not without effect on the share of resources: water, land, labor, and even ideas about the environment. Tourism affects many daily local aspects: first, use of scarce water, to be used by hotels or to embellish the scenery, then labor, diverted from agriculture, and even the definition of the oasis—a place for work or a place for a stroll?

I suggested earlier the notion of "socioecological landscape"; I have elsewhere discussed "socioecological resources" that form and define landscapes as being "classical," "instrumental," and/or "relativist."[2] The tourist modality in the Jerid dates to the Orientalist infatuation; tourism was first a luxury or adventure enterprise (at the end of the nineteenth century and beginning of the twentieth) before turning to mass tourism. In 1922, Le Grand Hôtel de l'Oasis was built in Tozeur, the biggest oasis of the Jerid; since 1980, Tozeur has had an international airport. The preoccupation and modalities shifted from the picturesque and ethnocentric to the pursuit of authenticity and cultural relativism.

Strangely, the Edenic idea of the oasian landscape that minimizes its labor component prevails among European middle-class tourists. Facing or inside the palm

Tourists in the Nefta palm grove

grove, they do not see gardens, or cultivated small plots, but a forest—a forest of palm trees. This can be partly explained by a European-centered historic mechanism: Europeans have their own history of the landscape and of "natural nature." According to Jean-Claude Chamboredon,

> The genesis of the countryside as an idyllic social setting results from a long process of progressive disappearance of the rural proletariat . . . since the second part of the 19th century. . . . The countryside can be perceived as a natural space after the accomplishment of the process of neutralization (depoliticization and homogenization) that erases the social oppositions and historic contradictions that are embodied in its spatial organization and practices. Then, the "de-socialized nature" can seem the place of a life subjected to natural rhythms, the haven of a traditional civilization, the decor of a direct contact with a transcendence (aesthetic or religious).[3]

Because European tourists of the oases do not see the obvious (to local eyes) social inequity in the resource division (land, water, workforce) embodied in the palm groves, these agricultural lands can turn into nature and "naturalized landscape."

We left the idea of defining the landscape to focus on the identity of the actors invoking the landscape. In doing so, we noticed that "landscape" is a congruent notion with the tourist practices of the oasian spaces:

> *Landscape* is the product of the view of someone who is foreigner to it. Man doesn't think of elaborating a landscaped representation of the place to which he is attached and where he works or lives. He maintains that space, preserves it and submits it to an order already registered in his mind. He decorates that space and embellishes it. But the conceptual schemes that guide his view. . .and that insert a value judgment in his visual analysis do not produce a mental click that instantly transforms a place into a landscape.[4]

Does this mean that the concept of landscape does not exist in cultures other than Western ones? We should once more reframe the question. Indeed, it is less important to know if a space is perceived as a landscape or not than to understand what is regarded as valuable in these spaces.

Gardens in the Jerid are tidied-up places, carefully designed. In one place we find beds of cultivation, irrigation networks, sometimes a sheep barn, and often the hut of the gardener. The garden is often small and is private property. It is enclosed by a fence of palms, to prevent people outside from seeing the intimacy of this "domestic nature." Tourists retain the idea that they are in a "forest" when strolling along the main paths of the palm grove (always the same paths, as they aren't comfortable on smaller ones because they feel that they are violating the locals' privacy).

Tourists often agree with national (and sometimes international) agricultural service officials. For example, both would call the agricultural landscape of the oases "traditional." That understanding has a positive value for tourists: they travel less to see exotic places than to experience authentic "exotic cultures." For the Tunisian agriculture engineer, this traditional quality has a more a negative value, and he's paid to change it: Jerid oases should take part in the modern agriculture project and provide currencies by exporting to Europe the most prized dates, the *deglet en-nûr* variety.

The agricultural services did not succeed in reforming old palm groves despite extended efforts. But, using the colonial pattern, they succeeded in creating ex nihilo palm groves of *deglet en-nûr* date palms. Why did they fail with the old palm groves? I think it is first because of misunderstanding of the function of gardens, the practices in the spaces of the palm groves, and the nature of this landscape.

OASIAN LANDSCAPES: FOR WHICH INDIGENOUS PURPOSES?

I used the phrase "domestic nature" to designate gardens. It is not only to underline the anthropic quality of the garden but also to emphasize the living, inhabited quality of these spaces. More than a workplace, these are an inherited patrimony for many generations, a place where gardeners spend the day even if there is no necessity of working. It is the place where gardeners express their aesthetic, their idea of beauty.

These gardens are invested of an almost Epicurean aesthetics, wholly in keeping with the Arab traditional garden. The five senses are engaged: watermelons in summer or broad beans in winter, the perfume of roses and jasmine, color touches amid the profusion of green elements, birdsongs, songs of the gardeners climbing the palm trees during pollination or in the evening, drinking palm wine. Ambiance is one essential component of aesthetics. An aesthetics that included any more than the visual appreciation of an environment always prompted incomprehension in agronomist engineers, both Tunisian and European, who, facing these gardens, saw only a vegetal confusion that was necessary to reform. The aesthetic quality of the oasian gardens, however, partakes of more than what is seen. This is the limitation of the use of the notion of "landscape"—a concept that usually rests on a visual definition ("all the visible features of an area of countryside or land," according to the *New Oxford American Dictionary,* 2d edition).

For the local gardeners, living off their plots in the old palm groves, to appreciate the garden is to share it. The garden is a center where constructions of space surround the gardener. The garden is an intimate center and nevertheless a communal space. This conception is not exclusive to the Jerid; one finds it, for example, "in the Japanese culture, [where] the appreciation of the beautiful sites is inseparable from human commerce (exchange of poems, banquets, tea ceremonies)."[5] Perhaps between the Jeridi gardener and the tourist, one identifies the opposition that Augustin Berque made between "the sociable landscape" and the "solitary contemporary landscape" inspired by Romantic literature and painting.

The oasian garden is private property, but also a place for collective presence: working together in long and difficult labor, and in idle moments, sipping tea during the day or drinking alcoholic beverages from the palm sap at night—talking, sharing experiences, stories, songs, news. In the oases of Jerid, one can always find men

An oasis garden

gathered in "*halga*" or "*ga' ada*." The *halga* term means "circle" in the local dialect and comes from *halaqa* in literary Arabic (when a sheikh teaches his disciples), and *ga' ada* refers to the seated position (dialect verb *ga' ad*, "to sit down"). During these moments, men tell the stories of *'antriya* (the tale of 'Antara—"Anter" locally, pre-Islamic poet and warrior), but it is especially a time when things can be said about politics, the police, and women. These masculine meetings are certainly occasion of verbal transgression, but also of exchanges of knowledge. The aesthetic standards are communicated, the collective sanctions the individual, knowledge is transmitted: tests and comments are exchanged in the common agricultural context; one listens, one remembers accounts and local stories. A part of collective existence takes place there, in the garden.

According to Jeridi tradition, gardens of palm groves are thus not only productive areas but recreational ones as well. The palm-grove gardens produce more than a biomass: they produce social fabric. In contrast, the modern technicist paradigm—represented, for instance, by agricultural engineers—views the oasian nature as an object of technical exploitation. The "salvation" of the agricultural future of the region, from this perspective, will be indebted to the importation of modern technical support: drills, tractors, and other technologies. This movement brings along other ways of thinking and practicing in the environment. If the oasian nature is an object of technical exploitation, building this object is possible insofar as the oasis is seen as a resource to be appropriated. Nature is to be colonized, desert to be fertilized: this constant "exteriority" toward the object legitimizes the scientific study, presumed to be the only means of rational exploitation.

BUILT NATURE: TOURISM IS SHAPING LANDSCAPE

What is true with the obvious intrusion of modern agriculture is true with tourism as well: the oasis represents scenery, an exotic landscape. The practices of modern agricultural and tourist actors make them keep their distance from the oasian space, which is a lived space for its inhabitants. This constant distance or exteriority toward the desired object legitimizes neither their colonization nor the scientific study, but a Romantic posture with respect to nature and a relativist reading with respect to local communities. For tourists, the exploitation of nature is less technical than ideological, but the effects are nevertheless considerable.

With a posture of distance, oases are islands in a mineral ocean. Maritime imagery is frequently used in dealing with the desert. The oases would be islands or ports, the desert a sea of sand, and camels the desert vessels. This is a particularly dominant metaphor in the Western imagination; many tourists are disappointed not to find oases always immersed in the sands, although sandy zones represent only one seventh of the Sahara.

Displaying the outward dimensions of oases is no longer sufficient; tourists demand at least functionalist explanations of "how it works," of the natural and social machinery that operates behind the décor. However, the landscape remains two-dimensional, guarding its third dimension: local daily life and its details. In the Jerid, a popular tourist attraction is a ride in a hot-air balloon, offering the possibility of a comprehensive view that remains exotic. In Siwa (Egypt), where a similar bird's-eye view is achieved by climbing a mountain that overlooks the oasis, I asked a local if enjoyed such a viewpoint; he replied, "Why should I go to see from above things I already know from below?"[6] This is enough to shake the universality of the concept of landscape. Nevertheless, this contrast between local peoples' and tourists' ways of perceiving the oasian environment and qualifying spaces does not ensure that the local environment or local interactions with nature will not change.

Males meeting to drink palm wine in the gardens under the palm dates

Tourist actors do not come to intentionally transform the oasian landscape in its material dimension—on the contrary, their first concern is to engage with an idealized past in the desert, a remote harmony with nature. Nonetheless, they transform oasian landscape in two ways: directly by spreading new ways to think about the relation to nature, and indirectly through the services and infrastructures organized to welcome them.

All facets of the local economy are intimately interlinked with the environment. Let us approach this issue by dealing with water, a key factor of life in the oasis. The competition for water is obvious: water is needed for agriculture, but is also needed for the tourism sector. Tourism service officers and hotel architects align their view with international standards and maintain that a swimming pool is required to satisfy European tourist expectations. In the Jerid, the mayor of Tozeur (the capital of the region) and owner of Dar Chraîet hotel (bearing his name) went even further: he offered a 50-hectare golf course in the desert. Along with thalassotherapy and cultural tourism, golf is part of the new measures adopted by the Tunisian government to diversify the tourist product.

According to the Tozeur golf website, the course "*overlooks* Tozeur's palm grove which is one of the most beautiful in Tunisia. And its 25 hectares of greens are irrigated with *recycled waters* to preserve the water table" (emphasis mine). In two sentences, you have two major imperatives toward local nature: a posture of domination (overlook it), and, for the sake of tourists' consciences, a response to ecological concern (recycle water). We can easily imagine that this recycled water could have been used for agriculture, which is suffering from lack of water.

Even if tourist guidebooks still mention, "over 200 springs" (as in the last edition of *Lonely Planet Guide*), no springs naturally irrigate the oasian gardens in the Jerid. Instead, deep drillings in the ground are used. They were dug first by French colonial farmers and administrators to free themselves of the complex indigenous negotiations for this resource, and then by the Tunisian Ministry of Agriculture. It was a progressive evolution over the last fifty years from "natural" springs, maintained by the local workforce, to water drilling maintained by the administration.

To say that the exploitation of water is a "mining" operation is not inaccurate, as the exploited resources are nonrenewable. Tube wells draw from the deep water table, which is insufficiently renewed.[7] Nobody knows if the water reserve will last ten or fifty years. The dried-up springs were compensated for by more drilling in the deep water table. Because of continuous depletion, those artesian drillings became increasingly less effective, and more energy (for motor pumps) is needed to draw the same quantity of water.[8] Deep drilling has removed the capacity for the old oases to live on natural springs. Thus there was no true profit: the drilling provided a temporary increase of surface area for agriculture, but now the upkeep of these new plots, and even the maintenance of the old ones, is increasingly difficult.

How can an economy of tourism be inserted in such a difficult context? Ironically, by building a golf course and encouraging the development of more hotels with swimming pools. In Tozeur, eighteen hotels and five residences offer 3,500 beds. The government aimed to ease congestion of the littoral sea-and-sun tourism with the construction of Tozeur's international airport.

NEW IDEAS FOR LOCAL RELATIONS TO THE ENVIRONMENT

Tourism shapes oasian lands indirectly through services and infrastructure, but also through the concept of landscape. Tourism offers new ways to "read" nature. We saw that "The 'landscape' is the product of the view of someone who is foreign to it."[9] The local importance of ideas brought in the tourists' baggage cannot be gauged by the amount of time spent in the region (in Tozeur, tourist hub of the Jerid, it is only one-and-a-half days per person on average—partially because of the difficult weather). Nor it can be gauged by the mere presence of tourists, this mobile mass of foreigners. The gap between visitors and the local population is designed by the local society to preserve itself from uncontrolled contacts. Some informal interactions are tolerated, often represented by the class of young candidates to emigration called "*beznesa.*" The government also reinforces the gap, to maintain both strong political control over the Tunisian population and the benefits of a tourist open-door policy (but it drives out the *beznesa,* to privilege more official interactions). The result of these two confluent strategies is a strange kind of tourist apartheid, where contacts are motivated mainly by business, sex, and emigration.

However, borders separate as much as they bring together, according to one of the famous paradoxes of Edgar Morin,[10] and interactions of the local and the global are always more complex than simple "culture shock": local communities haltingly give evidence of their desire to partake of the contemporary world, of the "global circulation of material and cultural flows." There is a perceptible "development of aesthetics of the '*furja*', of the panorama and the beautiful sight."[11] This aesthetic feeling that the space confers is only visual. It comes with neither a physical relation to the place nor with sociability. This local variant of aesthetics is implemented mainly for remote landscapes, especially displayed via television. The media is one way to hold the object at bay. This abstraction of the environment can be called "*taby'a*" (nature), a concept apart from the daily experiences of oasian people of the Jerid. Today, the young *beznes* vigorously refuse to set foot in the pepper beds of their fathers, but they will defend the oasis and argue for the safeguarding of its aesthetics.

Here is an example: to address farmers' complaints about the lack of water for irrigation, in 1996 the agricultural administration undertook a project to pave the beds of the main *wadi* of the Tozeur palm grove to reduce losses due to infiltration. Young people experienced this event very negatively. "The scenery is spoiled," they said. "The Government should have done it only in the hidden places of the oasis"— meaning the places where tourists do not go. This criticism relates to the loss of its "authentic quality" (which usually merges "traditional" and "old"). These *beznes* stand up for the "landscape" object, as it brings the tourists, of course—in which they have a vested interest. However, they also develop a real feeling for the need to safeguard an inheritance. This preoccupation with the environment or the landscape does not arise from the local categories of environment perceptions or practices. This operation was made possible only by the exteriorization of the oasian object for local actors supposed to be from the "inside." First, they have had to imagine which representation of "nature" the foreigners have. Palm dates are no longer, for them, just the most common and obvious cultivated plant (in local interiors, a contrasting representation is often displayed: big pictures of snowy Swiss mountain forests): no, they learned to behold date palms as exotic features. And substantial consequences emerge: some gardens undergo renovation. Where agriculture services failed to reform them, tourism succeeds, and some gardens are transformed into a camping site or a cafeteria.

The date harvest

In the same way, young people learned to not consider their local customs archaic (although many still want to emigrate to Europe, in part to escape from them), but as valuable components of the local cultural package—not to live it, but to display it. The tradition is a point of convergence between tourism and the local society: the 'adât wa taqâlîd ("local customs") are in both cases perceived as a heritage—fixed, timeless, and vouching for the true local oasian identity. In this respect, the local interface with tourism goes further in forestalling touristic desires: they even claim in the Jerid a local Berber heritage (a very recent claim) when such connection is unthinkable by the local community as a whole; to the contrary, local identity is built on Arab (and Muslim) affiliations. Tourists' demands (especially French and German tourists) to see Berbers is strong enough to be relayed and reified locally (French colonial discourse highlighted the Berber component of North Africa to legitimize its annexation to Europe, given the presumed Latinity of Berber culture). Concerning the palm-grove spaces, young people operate a "purification" by excluding the compromising practices—living the oasis from the inside, like their forefathers—and by bringing out only its aesthetics.

The Jerid municipalities (Tozeur, Nefta, El-Hamma, etc.) will probably accentuate this patrimonialization of the agricultural land in support of tourism. Despite the water deficiency for agriculture, drilling was undertaken in the palm grove of Nefta to put water in the dry beds of the main *wadi*. In view of the expected rate of flow, water will have infiltrated and evaporated before reaching the gardens. The goal, however, was rather a "visual and sound reenchantment" of the oasis—to restore its picturesque attributes according to presumed tourist expectations. The municipality has the same schizophrenic duties as the state: to take care of two poles of activity—agriculture and tourism—even if the most effort goes clearly toward tourism.

This concern is also clear in the new strategy given to the agriculture administration: preserve the "oasian aspect." A recent official tendency is interest in conservation of biodiversity, after having long encouraged the monoculture of the *deglet en-nûr*. Plantations organized by the state have been directed to have no more than 70 percent of groves in *deglet en-nûr*. The objective is to try to preserve the diversity of the Tunisian cultivars, in case the catastrophe scenario comes true: the *Bayoud* (a cryptogamic disease that ravaged Moroccan and Algerian palm groves) comes across the Algerian border and devastates the Tunisian oases. Therefore one can preserve at least the dominant visual aspect of oases even without an (exportable) date production.

Beyond the natural landscape, tourism has an efficient role in the arrangement of settlements. In Tozeur, the municipality decided that the local mud brick is the "authentic" feature of the traditional settlement; al-Hadawif, the most remarkable area in the old town, is actually made of local yellow mud bricks, but other parts of the old town, even older, are made using a different technique.[12] But the best way to promote tourism is to define the simplest visual identity, as in marketing: the mud brick has been widely spread everywhere in the tourist town circuit, to the extent that even some mud-brick walls were built to hide nonconforming architectures along main pathways. An architect, as early as 1991, was able to talk about a "folklorization of the space" and emphasized later the refusal of the inhabitants to get involved in such an "expression type" of their local identity.[13]

In Egypt, nearly the same process occurs in the Siwa oasis. In two decades, Siwa witnessed significant changes to the organization of its habitat.[14] While a radical change of space distribution of domestic units took place, the materials changed from a salty mud-clay mortar to the squared calcareous gypsum, inducing a fast evolution in techniques of construction. Although these two materials are both of local origin and production, the shift from one to the other is not without impact on the design of the habitat and the social uses of the dwellings. The argillaceous material seems to have regained favor by (political and commercial) promoters of the "traditional." The salty mud has been chosen to give a visual identity to Siwa, as a marker of the "traditional" for tourism purposes.

* * *

Reversing the usual process—by accommodating the reality to the copy—is not a new idea. Philippe Descola noticed that two tales of Edgar Allan Poe present this same reversal: *The Domain of Arnheim* and *Landor's Cottage*.[15] In these two stories, protagonists discover a piece of nature (a large domain in the first, a secret valley in the state of New York in the second) and dedicate themselves to landscaping their environment, to smoothing and transforming it until it becomes a "real landscape"—a succession of scenes, bearing the illusion of a Romantic painting.

Oases are artificial landscapes shaped by local people and environment conditions, but they are also shaped by external ideas, values, and expectations, and in this respect tourism has had a major impact, especially in recent years. The landscapes desired by tourism, compatible with the icon of the timeless "oasis," become increasingly a marketing feature; thus tourism is shaping the oasis landscape in its image.

NOTES

1   Vincent Battesti, *Jardins au désert, Évolution des pratiques et savoirs oasiens: Jérid tunisien* (Paris: Éditions IRD, 2005); and "De l'habitation aux pieds d'argile, des vicissitudes des matériaux et techniques de construction à Siwa (Égypte)," *Journal des Africanistes* 76: 1—Sahara: identités et mutations sociales en objets (2006), 165–185.

2   Vincent Battesti, "Les oasis du Jérid, des ressources naturelles et idéelles," in Michel Picouët, Mongi Sghaier, Daniel Genin, A. Abaab, Henri Guillaume, and M. Elloumi, eds., *Environnement et sociétés rurales en mutation: Approches alternatives* (Paris: Éditions IRD, 2004), 201–214; and *Jardins au désert*.

3   Jean-Claude Chamboredon, "La 'naturalisation' de la campagne: une autre manière de cultiver les 'simples'?" in Anne Cadoret, ed., *Protection de la nature: histoire et idéologie, De la nature à l'environnement* (Paris: Éditions l'Harmattan, 1985), 138–151; 142.

4   Gérard Lenclud, "L'ethnologie et le paysage, Questions sans réponses," in Claudie Voisenat, ed., *Paysage au pluriel, Pour une approche ethnologique des paysages* (Paris: Maison des sciences de l'homme; Mission du Patrimoine Ethnologique, 1995), 5–17; 14.

5   Augustin Berque, 1993.

6   Vincent Battesti, " 'Pourquoi j'irais voir d'en haut ce que je connais déjà d'en bas?' Comprendre l'usage des espaces dans l'oasis de Siwa," in Vincent Battesti and Nicolas Puig, eds., *Terrains d'Égypte, anthropologies contemporaines* (Cairo: CEDEJ, 2006b), 139–179.

7   A. Mamou, "Développement des zones sahariennes en Tunisie et son incidence sur les ressources en eau," in *Les oasis au Maghreb, Mise en valeur et développement, Actes du séminaire Gabès, 4, 5 et 6 novembre 1994* (Tunis: Cahier du C.E.R.E.S., 1995), 71–86; 73.

8   Michæl Richter, "Les oasis du Maghreb: typologie et problèmes agro-écologiques," in *Les oasis au Maghreb*, 29–56; 42.

9   Lenclud, "L'ethnologie et le paysage, Questions sans réponses," 14.

10   Edgar Morin, *Le paradigme perdu: la nature humaine* (Paris: Éditions du Seuil, 1973).

11   Nicolas Puig, "Entre souqs et musées: Territoires touristiques et sociétés oasiennes à Tozeur en Tunisie," *Espaces et sociétés, tourisme en villes* 100 (2000), 57–80.

12   Ibid.

13   Farid Abachi, *Les banlieues perdues ou la ville enfouie: Tozeur* (Tunis: Thèse d'architecture, Institut Technologique d'Art, d'Architecture et d'Urbanisme de Tunis, 1991), 139.

14   Battesti, "De l'habitation aux pieds d'argile, des vicissitudes des matériaux et techniques de construction à Siwa (Égypte)."

15   Philippe Descola, "Postface: Les coulisses de la nature," in Adel Selmi and Vincent Hirtzel, eds., *Gouverner la nature* (Paris: L'Herne, 2007), 123–127. Henri Duveyrier, *La Tunisie* (Paris: Librairie Hachette et Cie, 1881). René Pottier, *Au pays du voile bleu* (Paris: Nouvelles éd. latines, Fernand Sorlot, 1945).

CHRIS JOHNSON

9

DESERT TOURISM AS A VEHICLE
FOR NATURE CONSERVATION:
THE JORDANIAN EXPERIENCE

Jordan is a small, arid country, with a remarkable diversity of desert ecosystems, from extensive plains of basalt gravels and rocks, known as *Hamada*, to classical sand-dune systems and dry mountain ranges. They include the grand desert of Wadi Rum, with its monolithic towering cliffs of sandstone—internationally famous as a tourist site and as the dramatic backdrop to sequences in the classic film *Lawrence of Arabia*. Along the western flank of the kingdom runs a mountainous escarpment forming one edge of the Jordan Rift Valley, where the desert landscape gradually changes with altitude to dry steppe and open Mediterranean forest. Some areas along this escarpment, where desert vegetation grades into more moisture-loving plant communities, are rich in biodiversity and continue to support scarce large mammals such as wolves, hyenas, and ibex.

The protection and management of Jordan's special ecosystems is entrusted to a nongovernmental organization, the Royal Society for the Conservation of Nature (RSCN). It is one of the few NGOs in the world given such a national mandate by government. The society was formed in 1966 by a group of disillusioned hunters, worried about dwindling populations of game animals, and was the earliest voluntary conservation organization established in the Middle East. It started its first conservation program in the Eastern Desert, with the creation of the Shaumari Nature Reserve as a release site for the critically endangered Arabian oryx, and is now responsible for six protected areas, covering a land area of more than 700 square kilometers. These protected areas represent some of the finest landscapes and ecological systems in the kingdom and vary in size from the desert oasis of Azraq, at 23 square kilometers, to the mountainous wild lands of the Dana Nature Reserve, at 320 square kilometers.

Jordan has a wide variety of spectacular desert landscapes. Top, the "Wadi of the Teeth," Eastern Desert near Azraq. Bottom, the Dana Nature Reserve in the southern mountains of the Jordan Rift Valley.

For several decades, RSCN managed its protected areas as isolated sanctuaries, fenced and guarded from the general public and with little involvement of local communities.[1] This all changed in 1992 with the Rio Summit and the Biodiversity Convention. As a signatory of the Convention, Jordan was the first country in the Middle East to be awarded a multimillion-dollar pilot project under the Global Environment Facility (GEF) to develop a regional model of integrated conservation and development. The project was focused on the Dana Nature Reserve in southern Jordan, and its emphasis on linking protected area management to the socioeconomic development of local people ushered in a new era in conservation thinking, which RSCN continues to advance today. In particular, it kick-started the development of ecotourism as a principal tool for sustaining the management of protected areas and for engaging local communities in nature-based livelihoods. Since 1994, when the project began, RSCN has been developing ecotourism ventures in all of its protected areas, using the following definition of ecotourism as its guiding reference: *"Responsible travel to natural areas that conserves the environment and improves the well-being of local people" (International Eco-tourism Society, 1990).* Today RSCN is regarded as a significant player in Jordan's tourism industry, and its tourism operations are making a major contribution to covering the running costs of protected areas, despite the fact that Jordan sits in one of the most politically volatile regions of the world and has suffered from major oscillations in visitor numbers with each political crisis.

THE DANA STORY

The evolution of RSCN's approach to tourism begins in the scenically spectacular Dana Nature Reserve, just north of the famous archeological site of Petra. Here at the edge of the Jordan Rift Valley, an array of ecotourism facilities and activities were developed that attract more than 21,000 visitors a year and provide enough revenue to cover most of the reserve's running costs. Dana is widely considered a regional model of sustainable biodiversity conservation, and RSCN has been applying "Dana principles" to all of its protected areas.

Several thousand people from nomadic and settled communities live in and around the reserve, and many are partially or entirely dependent on the reserve for their livelihood. Their use of the reserve, however, is causing serious ecological problems, stemming from excessive livestock grazing, hunting, and fuel-wood collection. These activities had little detrimental impact in the past, when tribal populations were smaller and Bedouins could practice their nomadic lifestyles without issues of national borders, settlement policies, and major infrastructure developments. But now the pressure of these traditional livelihoods is concentrated in the last remaining areas of unspoiled landscape and is the major threat to their ecological integrity. In an attempt to solve this problem in the Dana and create more sustainable income sources, an ecotourism operation was gradually developed under the GEF project, building on the scenic, ecological, and cultural assets of the protected area.

The tourism concept for Dana was built on the premise that nature should come first, and local people should be the prime beneficiaries. The first stage in realizing this concept was to develop a zoning scheme, identifying areas where tourism (and other) activities can be accommodated with relatively little serious environmental impact. There are currently three major zoning categories: intensive use, semi-intensive use, and core zones. The first two categories are where all tourism activity is concentrated, and the core zones are where conservation needs are given priority and human access is restricted to enforcement rangers and occasional visits by researchers and management staff. The zone boundaries were derived from an analysis of baseline surveys of flora and fauna, including the known whereabouts of endangered species, and of existing patterns of local community use. Although biodiversity conservation is a primary objective of the reserve, the zoning scheme now in place represents a pragmatic compromise between conservation priorities and local community needs.

Dana Zoning Plan: Core zones, A to D—no public access, conservation given total priority.
Semi-intensive use zones, A and B—controlled public access, user limits, maintenance
of traditional livelihoods. Intensive use zone—main area for development and public access,
still with user limits, traditional uses maintained.

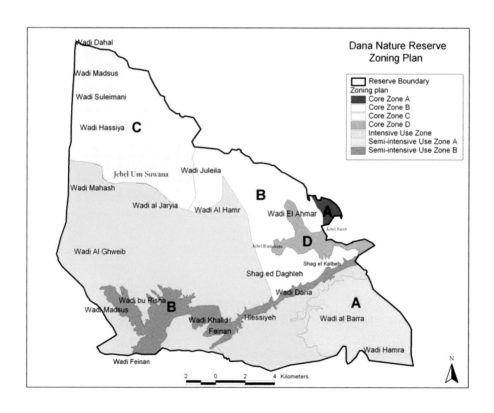

FROM CAMPSITE TO GUESTHOUSE

Following the introduction of the zoning scheme, a range of visitor facilities were constructed over several years, including a campsite, visitor center, hiking trails, and a strikingly designed guesthouse. The campsite was the first to be developed, as a small camping area had already been created before the GEF project started, although it was discovered from the baseline surveys that the area fell into the core zone designation. The camp, however, was already known and used, and was originally a favored Bedouin camping area, occupied every summer for goat grazing. Clear topographical boundaries were therefore defined for the campsite, and all hiking routes out of the site were directed into the semi-intensive and intensive use zones. A shuttle-bus system was also introduced to ferry visitors from the reserve entrance gate to the campsite, eliminating access by private vehicles. With the shuttle system in place, number limits were then defined for visitors entering the reserve and for the trails and campsites, based on estimated carrying capacities and a simple monitoring program devised to assess visitor impacts.

The success of the campsite in attracting visitors prompted the creation of a small guesthouse close to the edge of Dana Village. Originally intended as accommodation for researchers under the GEF project brief, the building was appropriated halfway through construction and converted into a nine-room lodge for tourists. (The researchers were eventually housed in a restored village house.) This decision reflected not only a growing awareness of the potential of tourism in Dana to be a key conservation manage-

The Dana Reserve tourist facilities: top left, the guesthouse; top right, the campsite kitchen and store; bottom left, the "Dana shuttle" taking visitors to the camp (one of the access controls); bottom right, the campsite tents against the mountain backdrop.

ment tool but also the dramatic location of the building, perched at the edge of the precipitous cliffs of Wadi Dana and close to an old, scenic village. The architect had also designed a simple building, merging elements of village vernacular architecture with more modern features, to create a building with a definable Arab character that was nonetheless unlike any other in Jordan. It was soon apparent that this mixture of good design, location, and exceptional visual and cultural interest was a winning combination for attracting tourists, and the Dana guesthouse became—and remains—one of the most popular and viable tourist facilities in the nature reserve.

LOCAL PEOPLE ON BOARD

Throughout the development of these tourism and conservation initiatives, local people were part of the process, and an extensive capacity-building program was undertaken to enable them to share in the benefits. Employment opportunities were one of their chief concerns, and RSCN introduced a strict policy that all tourism jobs were to be filled by local residents. The organization also tried, wherever possible, to provide these opportunities to the families most dependent on the nature reserve for livestock grazing and other subsistence livelihoods. A range of small businesses were also developed to produce tourist souvenirs and other products, providing additional employment and revenue, especially for the women of the communities.

When the project began in 1994, there were fewer than 100 annual visitors to Dana, but by the year 2000, this number had increased to almost 20,000.[2] The tourism and sales receipts over this period generated enough revenue to provide fifty-five full-time jobs along with direct and indirect benefits to more than 800 people in the surrounding communities. Moreover, by 1998, the level of revenue was such that all the running costs of the nature reserve were covered, making it financially self-sufficient. Because the reserve is large by Jordanian standards and of recognized global significance for biodiversity conservation, this is a remarkable achievement, for which Dana has won international awards for sustainable tourism.

Despite the Dana project's overall success, not all local people supported the nature reserve, and some remain openly hostile to its existence. At the start of the project there was significant resistance from local communities, largely because the reserve itself was established five years previously without consultation. This created a legacy of resentment and mistrust that was difficult to overcome. Only through continuous engagement with these communities, and the generation of tangible economic benefits from tourism and craft production, has RSCN gradually been able to change attitudes and create a constituency of support for the nature reserve. This lesson has been absorbed and institutionalized by RSCN, such that it now has a well-defined process of consultation and engagement with local communities prior to the establishment of any new protected area.

THE FIRST PURPOSE-BUILT ECO-LODGE

Encouraged by the benign impact of Dana's initial tourism enterprises, RSCN embarked on the construction of a purpose-built eco-lodge at the western gateway to the reserve during 2003. This was located in the remote Wadi Feynan, on the site of an old copper-mining research base, and built with funds provided by USAID. Distant from roads and power supplies, the lodge represented a brave attempt to create a unique tourism experience in Jordan and bring enhanced economic benefits to the Wadi Feynan Bedouins, among the most underprivileged tribal groups in the kingdom; their reliance on intensive goat grazing in the nature reserve is a cause of many ecological problems. Its location was also part of a strategic conservation initiative to use tourism to offset the threat of open-cast copper mining in the Feynan area. There is persistent pressure from government and private-sector companies to rework the extensive but low-grade copper deposits surrounding the lodge (which, ironically, provided the economic base of previous civilizations in the region, from Neolithic to Islamic); development of tourism provides a far more environmentally sustainable livelihood option.

The Feynan Lodge, like the Dana guesthouse, is an exceptional building. Taking influences from ancient caravanserai and Yemeni architecture, it provides twenty-six rooms, all organically shaped and different in layout. It incorporates several environment-friendly features including solar power, high insulation, and passive ventilation; in the absence of mains electricity, it is lit at night by candlelight, which gives a special atmosphere to the building and creates an unusual attraction for tourists. It was opened in September 2005 and provides direct employment for fourteen local Bedouins, as well as income for local service providers such as the village shuttle service, bread makers,

The Feynan Eco-Lodge.

vegetable growers, etc. By the end of 2006, it had already attracted sufficient visitors to cover its operational costs and make a small profit.

With its array of tourist facilities and activities, Dana is featured in several international tourist guides including the *Rough Guide, Insight Guide,* and *Lonely Planet,* but with its growing reputation and increasing visitor pressure, tight controls on the number, distribution, and behavior of visitors continue to be applied. So far the site is showing little sign of excessive negative impacts from tourism; on the contrary, routine monitoring programs show increases in the populations of several key species such as the Nubian ibex, griffon vulture, and Syrian wolf. Another notable aspect of Dana is that it attracts a high percentage of Jordanian visitors—more than 50 percent in most years—which both helps to maintain income levels when international tourism fluctuates and creates in-country support for protected areas.

INCOME DRIVERS AND FINANCIAL PERFORMANCE

The main income drivers operating in Dana are entrance fees, accommodation fees, and activity fees (mainly guided hikes). A summary of the financial performance for Dana in 2007 is given in the table and provides an indication of the source of revenues and overall balance sheet.

It is interesting to note that a relatively small number of visitors can generate enough revenue to make a major contribution to the running costs of a large protected area, if this revenue is channeled directly into the site. In 2005, the contribution was proportionally larger (92 percent break-even), but was affected by the opening of the Feynan Eco-Lodge in 2006 and the relatively small operational profit it made during its first and second years. Entrance fees make a relatively small contribution to running costs at this level of visitor use, with an average of $3 per person compared to the accommodation at an average $26 per person.[3] It should also be noted that construction of the Dana facilities, including the Feynan Lodge, was financed by international donors, and RSCN is not required to repay the initial capital outlay from revenue. While

SUMMARY OF FINANCIAL PERFORMANCE FROM DANA NATURE RESERVE, 2007

Number of bed nights	11,300
Total number of visitors (including day visitors)	21,200
Overall occupancy rate (guesthouse, campsite, and lodge)	54%
Total revenue from entrance, accommodation, and activity fees	$392,000
Total operating costs of the nature reserve	$513,000
Break-even point	75%

this confers a great financial advantage, RSCN regards the investment of donor money in tourism facilities as a major way of ensuring the sustainability of donor-supported conservation programs.

REPLICATING THE DANA APPROACH

Building on the experience gained from the Dana tourism programs, RSCN has been transferring the "Dana approach" to the other five protected areas under its jurisdiction. These areas are very different in size and landscape character, ranging from a restored oasis in the Eastern Desert to a large arid mountain system on the shores of the Dead Sea, deeply incised by wadis and flowing rivers. They also include two forest reserves in the north of the country that provide a great contrast to the desert areas of the south and east. Collectively, the RSCN has developed six permanent lodges/guesthouses, three campsites, three visitor centers, and many hiking/activity trails. The character of these facilities differs for each site, and great effort has been made to take inspiration in their design from the local landscape and human and cultural history. There is, for example, a recently completed lodge in the village of Azraq in the Eastern Desert, created from a renovated British army field hospital of the 1940s. The design carefully maintains the

Tourist facilities recently completed by RSCN, showing the wide range of architectural styles. Top left, wooden cabin in the Ajloun Forest Reserve; top right, silver-domed chalets at Mujib on the shores of the Dead Sea; below, Azraq Lodge in the Eastern Desert, a converted British military field hospital dating from the 1940s.

character of the original buildings, but creatively links them to a modern, sculptural new wing, incorporating tentlike structures to hint at their military past. At Mujib, on the shores of the Dead Sea, the chalets have domed roofs painted silver to mirror the color and shine of the surrounding seascape; while in Ajloun, wood has been used in the tented bungalows to reflect their forest setting.

With this portfolio of projects and experiences, RSCN is now confident about the benefits of desert tourism for nature conservation. With careful planning and management, tourism has been shown to provide an effective vehicle for:

- Generating substantial income for biodiversity protection.
- Creating jobs and revenue for local communities and thus more support from local people for conservation.
- Helping Jordanians to appreciate and value their natural heritage.
- Ensuring that conservation is demonstrating economic value that government officials and other decision makers recognize and appreciate.

WILD JORDAN

Although the benefits of tourism are evident, they have not been easily won. Jordan undoubtedly has some spectacular landscapes and a fascinating cultural history, but it also lies in one of the most politically unstable regions in the world and is affected by unpredictable and violent events on its borders. It became apparent to RSCN early on that good marketing was the key to making its tourism ventures viable and sustainable, especially as Jordan was not known as an eco- and nature tourism destination, but more for famous archeological sites such as Petra and Jerash. This realization may seem obvious, but understanding the value of marketing is rare in NGOs, and many well-intentioned tourism projects have floundered because little attention has been given to it.

Initially, RSCN created a tourism unit, embedded in the conservation department, with staff devoted to promotion and bookings. This worked well at the beginning, but as the number of tourist facilities increased, it was not able to generate enough impact in the marketplace. Apart from the lack of professional marketing staff in the unit, it was proving difficult to get local tour operators to support the new ecotourism opportunities being opened by RSCN. With their long history of conventional archeology tours, most Jordanian operators were not comfortable with the new product, having little understanding of the philosophy of ecotourism and what it entails.

Faced with this response from the industry, RSCN decided to become more entrepreneurial itself and developed the idea of "Wild Jordan"—a separately branded division devoted to the development and marketing of its ecotourism. This division recruited a team of marketing and PR staff from the private sector and introduced private-sector techniques of commissions and other performance-related incentives. Under the slogan "helping nature, helping people," the Wild Jordan division is proving increasingly effective in promoting RSCN's ecotourism products. It has cultivated the local tour operators through constant interaction and has secured agreements with more than thirty companies for bringing clients to RSCN facilities. It has also introduced several innovative marketing initiatives, of which the most pioneering is the Wild Jordan Center in the heart of the capital city of Amman. This purpose-built center, perched high above the old city, combines a tourist information focal point with a large nature shop and whole-food restaurant to provide a popular showcase for RSCN's tourism and handicraft products. Sales revenue contributes significantly to the annual conservation budget of RSCN.

ISSUES AND LESSONS LEARNED

The main problems affecting the development of desert tourism as a conservation tool in Jordan have been social and business-related, rather than ecological. As noted, key ecological indicators in the Dana Reserve, which is the most developed for tourism, have not shown a negative relationship with increasing tourism activity. The one exception concerns local community livelihoods. It was envisaged that increasing job opportunities in tourism would reduce the dependency of local people on ecologically damaging land-use practices, especially goat grazing. This does not, however, seem to be the case. A study conducted in Dana in 2001,[4] five years after tourism began, concluded that RSCN's socioeconomic strategy (including ecotourism) had not significantly reduced the level of goat grazing in the reserve. The author notes, however, that the strategy "has been very successful in improving the attitudes of the local population toward conservation and the presence of the reserve."

One of the most visible social impacts has resulted from local entrepreneurs capitalizing on the tourists attracted by RSCN. This is most evident in Dana Village, where

The Wild Jordan Center.

Dana Village, where local entrepreneurs have capitalized on the tourists attracted by the nature reserve, but with little regard for village history and architecture, which threatens to destroy the visual quality and social integrity of the Village.

resident cooperatives and business-oriented individuals have created small hotels from old village houses to exploit the growing reputation of the Dana Nature Reserve as a tourist destination. This in itself is not a problem—indeed it would normally be welcomed by RSCN—but the proprietors have shown little regard for the architectural, historical, and visual quality of the village, and the resulting hotels and guesthouses have become eyesores. Apart from unsympathetic restoration, intrusive signing, strings of neon lighting, and a host of small, kitschy interventions are seriously affecting the architectural integrity of this Ottoman village. The local cooperative hotel also started employing foreign women for housekeeping and waitressing, undermining the benefits of tourism for local employment. Developments such as these, as well as attempts to compete for customers, have led to disputes between owners and families, and to criticism of RSCN for restricting development inside the protected area. RSCN is now planning a major project to involve the Dana Village community in creating a restoration and tourism development scheme for the whole village, to create more equity in benefit sharing and bring architectural quality back to the village as a driver for better tourism returns.

Another issue has been pricing for Jordanians. RSCN has set its pricing policy for entrance, accommodation, and activity fees according to the need to cover all tourism operational costs and raise revenue for supporting protected area management. Although the fees are not high by international standards (the average rate for a double room is $60), they are expensive for a large proportion of the Jordanian population, and RSCN is sometimes criticized for "pricing out" sections of society. This is an important philosophical issue for an NGO that has been given responsibility for managing natural heritage sites on behalf of the nation, since access to these sites is arguably the birthright of all Jordanian citizens. Indeed, the whole idea of RSCN becoming business-like and commercial in the interests of conservation is not always received favorably. RSCN has tried to reduce the perceived inequalities in pricing by charging lower fees for Jordanians than for international visitors and is also planning to introduce free "open days" in all protected areas. As for bridging the philosophical divide between business and conservation, RSCN now makes sure that the conservation benefits of tourism are well promoted in its marketing materials, and it has created institutional mechanisms that encourage cooperation between the staffs of the business and conservation departments.

Although RSCN has made major progress in convincing local tour operators to support RSCN's ecotourism products, there remains an important and unaddressed concern: the weakness of Jordan's search-and-rescue system. Tour operators are reluctant to place their clients in remote locations without effective safety measures. Jordan does not have a coordinated national system, and rescue in the field relies on local coordination between the police, civil defense, and air force, which are often ill-equipped for dealing with remote location emergencies. RSCN has taken it upon itself to train its guides and field staff in the required skills, but some tour operators remain concerned

Search-and-rescue capability is still a major concern of tour operators. RSCN staff teams have been trained to handle most emergencies, but there is no coordinated national system.

that without a national backup system, their clients will be endangered and could have grounds for lawsuits in the case of badly managed accidents.

The last significant problem RSCN has faced in its tourism ventures is related to maintaining operational standards and design quality. All of RSCN's tourism staff are recruited locally as a matter of policy, and capacity building for these staff is now a routine function of RSCN's work. In general the staff performs very well, and Jordanian hospitality is legendary, but problems persist in maintaining operational standards. There is a tendency to let standards slide, particularly in relation to the small, detailed service requirements that make the difference between a satisfactory service level and an excellent one, or between a service and an "experience." Examples include maintaining agreed menu choices, such as having fruit at breakfast, as preferred by Europeans, providing room information to visitors, folding down sheets carefully, offering "touches" like pillow gifts, maintaining furniture layouts, and repairing damaged and worn items in guest rooms. Some important environmental issues also tend to get overlooked such as segregation of garbage and maintaining "backyard" tidiness. Much of this stems from the difficulty of finding key personnel with the right managerial approach, especially in remote locations. RSCN has experimented with bringing in hotel trained staff from outside to act as counterparts for local staff and develop capacity, but none have been willing to stay in rural locations for an adequate period of time. For this reason, the idea of forging partnerships with private-sector companies or developing private-sector concessions is now being considered.

FUTURE CHALLENGES

The major challenges facing RSCN's desert tourism model in the near future are ensuring the effective management of the existing portfolio of tourism facilities, such that they continue to attract enough visitors to cover operational costs and generate significant conservation benefits, and accommodating the expansion of tourism ventures into four new protected areas soon to be designated in the Jordan Rift Valley. In each case, RSCN's strategy will need to take into account: increasing competition in the ecotourism sector in Jordan and RSCN's capacity to manage more tourism initiatives without diluting its primary conservation role. The first private-sector plans for eco-lodges outside RSCN's protected areas emerged during 2007, and it is clear that RSCN will not have a monopoly on nature/eco- and adventure activities in Jordan for much longer. It is also clear that RSCN cannot expand its tourism operations simply by expanding its overstretched core staff. Already the number of staff involved in tourism and other socioeconomic programs threatens to overwhelm the number in other divisions, and the issue of finding the right people for managing on-site facilities in remote areas is likely to prove an even greater problem than it is at present.

With these challenges in mind, RSCN is planning to restructure its operations and create a much greater role for the private sector. It has a well-developed scheme for facilitating private investment in the construction of a five-star eco-lodge in a new protected area just south of Petra. This will be the first time a protected area in Jordan has been opened up for private investors, and it will be watched closely to ensure that it is able to meet conservation and socioeconomic objectives as well as commercial ones. The concessions system in U.S. national parks has been examined as a possible role model, although RSCN's requirement to have tourism enterprises supporting protected area management costs (in the absence of adequate government funding) does not apply to private-sector operators in the U.S. system. The distribution of returns from a Jordanian concessions system will therefore need to be structured quite differently. For this reason, RSCN is considering setting up its own eco-lodge management company in partnership with Jordanian entrepreneurs. This could be developed with a strong commercial base and an equally strong ethical operating philosophy. Whatever approach is eventually applied, the primary aim is to use the private sector and private-sector approaches in innovative ways to enable desert tourism in Jordan to build on the pioneering work of RSCN and become an even more effective vehicle for conserving the fragile biodiversity and spectacular landscapes of this desert kingdom.

NOTES

1 The exception was the Shaumari Nature Reserve, which had a small animal collection and was open to visitors and school groups.

2 These are people entering the reserve and paying entrance fees. Many visitors stay in and around Dana Village and other areas on the periphery and do not pay to enter. These are not recorded in statistics.

3 Entrance fees have since been raised to $5 for Jordanians and $10 for internationals.

4 Amberly Knight, "Combining Conservation and Development: An Evaluation of the Socio-economic Strategies of the Royal Society for the Conservation of Nature in Dana, Jordan," master's thesis, Brigham Young University, 2001.

REFERENCES

Johnson, C., and T. Hawa. 1999. "Local People Participation in Jordanian Protected Areas: Learning from our Mistakes."

Jordan Tourism Board. 2006. "Eco-Jordan."

Knight, Amberly. 2001. "Combining Conservation and Development: An Evaluation of the Socio-economic Strategies of the Royal Society for the Conservation of Nature in Dana, Jordan." Master's thesis, Brigham Young University.

Royal Society for the Conservation of Nature, Jordan. 1997. *Dana Management Plan*, first edition.

Royal Society for the Conservation of Nature, Jordan. "Wild Jordan" website: www.rscn.org.jo

GINI LEE

10

THE INTENTION TO NOTICE:
EXPLORATIONS THROUGH
EPHEMERAL DESERT LANDSCAPES

The structure is obvious in its otherness;
even from a distance, the twisted, folded, rusted, discarded iron captured
and re-formed around the detritus of a tree stump.
Chance and abandonment has caused this material to reach here;
the momentary force of
a long-past flood has made the wrapping possible.
The surface still exhibits the corrugations of manufacture, yet this facsimile of wall,
the material of enclosure, is now transformed into a facsimile of a miniaturized landscape,
an eerie echo of the timeless folding of
the geology of this arid place.[1]

The annotated illustration at left is an account of only one of many explorations into ways to uncover and then to respond (through design) to the intangible and mutable qualities of the constructed and natural landscapes of the desert. My explorations are instigated by invitations to notice and then to intervene in [extra]ordinary landscapes, with a curatorial intent that seeks to encourage creative investigations into desert places through the practices of touring and collecting, leading to archiving. My method is to make itineraries framed around the ephemeral conditions noticed across a number of sites in the South Australian outback. And my curatorial design approach effects minimal intervention and encourages a facilitative community of practice, one that draws on chance encounters in desert architectures and landscapes leading to recording traces and events, large and small, material and immaterial, to encourage a physical and a conjectural archive. I take a range of elements, ground conditions, and processes and investigate them individually and from various relational perspectives. The tour operates within performative and temporal space; the collection is the material that is worked with; and the projects are positioned in the cultural, ecological, and political structures that preside in ordinary desert landscapes. This curatorial methodology is also framed by the ethics and aesthetics of postproduction theory,[2] where the essential attributes of such a design practice embrace:

The transformational: of familiar, often mass-produced or discarded, unused objects and situations, as devices that contribute to new scenarios through processes of collaboration, participation, installation, performance, and the development of new strategies for intervention into everyday spaces and landscapes.

The participatory: where passive viewers and consumers are encouraged into active collaborations within strangely familiar circumstances, which are undertaken in altered situations and discourses toward the creation of new communities, albeit ones that may exist only temporarily.

CONCEPTS OF DESERT AND ARID: LOCATING ORATUNGA

The places described in the following work, however, do not necessarily appear as normally recognized typical desert landscapes; they are often richly vegetated and support a range of pastoral and other practices. To gain understanding of my (Australian) desert context, I propose a working definition of what it means to understand/approach the notion of desert. Dictionary references suggest that deserts are areas of land, usually in very hot climates, that consist only of sand, gravel, or rock with little or no vegetation, no permanent bodies of water, and erratic rainfall. They are places or situations that are devoid of some desirable thing—devoid of life, wild, bleak, and uncultivated. (Here I find it particularly interesting to note the concept that deserts are both physical and situational, as if to suggest a performative framing.) Although much of central Australia is understood as a desert(ed) landscape, it is also most known as the Outback or the Bush, described as arid—a region in which annual rainfall is less than 25cm—making it, to many, completely lacking in interest or excitement. As we who work in deserts/arid lands know, however, this lack is not ever the case.

My investigations have landed specifically at Oratunga, a 75-square-kilometer pastoral lease in the arid high country of the North Flinders Ranges in South Australia. It exhibits histories of vast geological and vegetative change, aboriginal occupation over thousands of years, and practices of exploration and mining, pastoralism and tourism, art and recreation. It had been in drought for eleven years when I arrive, but recently there had been some rain. I was told that Oratunga translates as "the unknown."

When I decided to work with the arid outback and with Oratunga as a specific place, my guiding idea was to evolve ecological practices of management and stewardship in marginal arid landscapes that are at the same time aesthetically and geologically important and unique. I imagined desert places as expanded design studios, where people could be encouraged to visit and act in ways that allowed for creative practices to be undertaken and reconciled with the fabric, culture, ecology, and politics of the Flinders Ranges and the broader region.

What are the possibilities for design practice between not-knowing (a place) and knowing (a place) when we travel through remote arid lands? To investigate these possibilities, I devise projects for developing tactics for interaction and intervention, through methods that involve touring and curating through noticing. Through examining my photographic records and notes made while traveling and noticing, the following four propositional situations emerge that underpin a curatorial design practice:

1. Layers of dissimilar data are reconfigured as interconnected collections. Artifacts/objects encountered and/or collected do not "fix" space; rather, they act as points or moments to move with/against.
2. Scale relationships dissolve where the landscape may be experienced through the curated tour, over the course of three hours, three days, or any other temporal interval.
3. Devices placed in the landscape invest in material collections and temporal spaces the agency to make many itineraries.
4. Interventions may not be permanently sited/fixed in place. An ephemeral museum/gallery is a conceptually nomadic expanding archive that is transferable from place to place.

In traveling through my arid landscapes, I adopt transformational and participatory practices to interrogate curation of an expanding archive. The common medium is the materiality of everyday environments where color and texture effects are transformed into tangible or imagined narratives of the traces of occupation and event. Noticing and recording such traces prompts embarking on new investigations toward re-presenting the Outback landscapes and urbanscapes that I pass through. I now elaborate on three scenarios for curatorial design, based on noticing while touring and curating; I present extracts from projects I have made, sometimes with others, over the course of eight years.

NOTICING WHILE TOURING AND CURATING IN THREE WAYS: ON LOCAL COLOR

This first account recalls recording small and remote country towns for the Placecards postcard project. At the time, my documentary intent was to capture the scenographic experience and aesthetic qualities of places in the process of change and/or abandonment, before the disappearance of the palimpsest of their creation through the wear and tear of their existence. I made a series of chorographically arranged works, juxtaposed in the sequence in which they were taken, independent of aesthetic or other thematic concerns. My images reveal the narratives of specific places through drawing out the particular material surface conditions of places.

e ground

mined from stone, gathered from occupations past and present

MINTABIE   SOUTH AUSTRALIA

handmade dwellings emerge from red and white ground
mined from stone,
gathered from occupation past and present

The postcard shown is an extract from the series of Placecards for twenty small South Australian settlements, a project that creates a color coding and color language for remote landscapes and towns.[3] Over two years I traveled and made a photographic collection of hundreds of images, with particular emphasis on the colors and textures of the urban and design fabric that I noticed. Evident in the surface conditions of the material and cultural fabric of these places is the layers, built up over only the past 150 years, of newly constructed urban landscapes superimposed over one of the oldest landscapes on earth. These towns were first made by the pioneers of the nineteenth century, but the changing operations and fortunes of the twentieth century have transformed their everyday character. I set out to look for examples where, over time, collisions of building types, styles, materials, and markings represent the transformative and cumulative histories of each place. Behind the material stories of these places, visual and oral narratives emerge, and every marked/made surface resonates with local cultural and political histories.

The Placecards project prompts ways of looking at a critical perspective on color theory based on a cultural framing of remote arid places. Both visual and oral narratives are the languages that translate the stories of these places to unfamiliar audiences. Every marked/made surface resonates with the cultural and political histories of places. My color palettes are characterized by more than sampling of the material and a subsequent chromatic arrangement. They include the literal markings of place, the mappings, the signposts, weathering, collection, and response to change that speak of a diverse and rich texture of arid color.

Subsequently I explored the making of a local color code informed by Italo Calvino's textual languages, deployed to make an expanded visual language. Appropriating his text structured around the intangible qualities of terms such as quickness, multiplicity, and lightness, I annotate my images with abstract textual impressions made during the time I was reworking my archive to make postcards. In demonstration, I reproduce below one of six critical writings around the making of a local code.

LIGHTNESS

Italo Calvino's thesis on lightness suggests a duality in the way we simultaneously perceive lightness and "a respect for weight." His narratives seek "a verbal texture that seems weightless," that embraces abstract descriptions to notice subtle and imperceptible elements through the presence of images of lightness that acquire emblematic and intangible value.[4] My method selects certain images to expand upon the concepts of lightness in the outback town of Marree, some 800 kilometers north of Adelaide—a place of harsh desert existences, complex cultural interfaces, and slow abandonment.

Early morning light in the Marree public park casts deep shadows on the stone paved
ground, the grey of the stone and concrete embedded in the red brown earth,
in stark contrast with the dusty surface on which the town rests insecurely.

This first Marree story works on a subtraction of weight and invites us to "gaze upon what can be revealed by indirect vision (which is) not about metaphor, rather the lessons lie in the detail of the literal narrative."[5] In this sense, spatial and temporal elements combine to convey the everyday language of the town in a moment that is simultaneously an outsider's experience of the place, captured in a chance encounter. Marree is insinuated into my personal history of changing circumstances, where a moment's lightness is characterized as visual immateriality. John Rajchman has written on lightness with regard to the "disappearance of the anchoring of place, region or proximity" that allows spatial and material abstractions to emerge.[6]

The old Ghan train tracks gash through the centre of town, contained now within the town boundary and serving only the relic hulk;
that of the stationary rusting, scarred and incomplete engine.
Abandoned after the final run, seemingly in mid-journey, because Marree is halfway, weathered red and white paint lightens towards transparency.
But there are other contemporary lives and messages transported by the train; now they are of the town, lives in white chalk layering weight over a slowly dissolving base.

In the train remains, materials and textures belong to haptic rather than optic space, where "the light abstraction of assemblages that take one out from the gravity of locales and regions, bases and heights, release another more disparate movement." [7] And the quality of lightness in practice is about lightness of touch and of intervention. Color mobility, made through weathering and wearing processes, allows a textural transformation of objects revealing a lightness of surface and the potential for "other" canvases upon which to mark a town.

In conceiving of a catalog of parts for a local or regional code, two actions are pertinent: first, the application of multiple narratives to the visual collection, and second, an ordering according to carefully considered themes based upon the knowing of a place. The functional, symbolic, subjective, and/or empirical knowledge gained through numerous readings of the material culture of places resides in the visual languages "written on the walls."

What is essential to the development of local codes, and of a thematic coding system, is an acknowledgment of the potential for a multiplicity of readings and arrangements for local environments. Beyond the scenographic, curatorial ordering and caretaking of local palettes also conceives of a program of enduring dispersal such as the postcard, the publication, the exhibition event, in order to move the local beyond the personal archive and into a contribution to the public realm.

NOTICING WHILE TOURING AND CURATING IN THREE WAYS: ON GUIDING AND MAKING ITINERARIES

The second account is a record of guided walks made as chance, explorative events. These performative travel stories involve collaborators moving slowly through landscapes, with the absolute intent to notice the immediate detail of rare and mythical encounters with past occupants of the landscape. The outcomes are framed only through opening up the chances of something partially known being rediscovered. This drifting traveling is reminiscent, although without overt political intent, of the attitudes of Guy Debord, who influenced a generation of walkers. His conceptual *dérive* exhorts engagement in a process where "play-ful constructive behaviour and awareness of psycho-geographical effects" allow travelers to "drop their usual motives for movement and action... and let themselves be drawn by the attractions of the terrain and the encounters they find there."[8]

In his writings on a provisional theory of Non-Sites, Robert Smithson proposes that to draw a plan or prepare a topographic map of a place or landscape is to make a two-dimensional analogy or metaphor; and further, that this logical picture rarely resembles the original. By expanding this action into making a three-dimensional logical picture, the actual site can be further abstracted to represent that site elsewhere—the making of a Non-Site. In visiting the Non-Site, Smithson advises that it is possible to undertake a fictitious journey, one that can be invented or artificial, in effect a non-trip to a Non-Site. Here one can travel in the "space of a metaphor" and "everything between the two sites could become physical metaphorical material devoid of natural meanings and realistic assumptions."[9] Smithson regarded this speculative theory as tentative and able to be abandoned at any time; and I seek to attend to this advice as necessary.

Smithson's proposal informs my speculation into a method for constructing journeys through what are generally regarded as wilderness landscapes. The sublime Flinders Ranges is a landscape of deserts, plains, and high rocky hills and plateaus, of dry and stony creek beds. It is also a landscape of ruined and functioning settlements and outstations, eroded streams and hillsides, of dusty roads, fence lines, and watering devices that reveal the presence of years of human reworking of what is now the cultivated remote landscape.

Walking the creek beds in the high country reveals an immense complexity of detail in the surface of the creek floor, interspersed with the detritus of past floods that exposes evidence of human occupation in otherwise uninhabited local landscapes. During my wanderings, I notice and record the transient moments where detritus and landscape meet and superimpose, and I try to weave a narrative of occupation through an itinerary that may be also a Non-Site—an itinerary that refers to places of other enterprises, a metaphoric reconstruction of the original moment and place that generated these fragments that now lie in the creek bed.

The subsequent arrangement of images as a series of views describes the event of material being formed through interaction with chance forces of nature. Each series moves between contextual detail and takes in the space where the material has been embedded; each moves into an absence or extraction of visual context, but reveals more of the forces and traces inherent in the material itself. The imagined narratives work across scale, in ways that oral stories work between description and action. The serial events uncovered through the journey along even one creek suggest for me the making of an ephemeral garden of sorts, where the underlying garden structure and planting is formed by the "natural" landscape; the materials that end up captured in the creek are the collections and artifacts that "humanize" and construct the landscape, and the maker is the flood.

The Oratunga landscape is one where a local walk is usually measured in time rather than distance and may be taken alone, but more often with friends or visitors. One heads off along a dry creek bed, along a sheep track or dirt road, or up a hill. Serendipity is your guide; landmarks and punctuations in the walk might be geological or topographical—a particular, peculiar outcrop on a hill, or turning a corner in the creek and coming across a rocky gorge or even a spring, in the dry landscape. The signs of past occupations are often found in surprising and remote places. Making up scenarios around twisted, rusting material, barely recognizable as the thing it once was, can make for a lengthy ramble. What follows now is the account of one such guided tour.

SOMEWHERE IN GLASS GORGE

I had heard that there was a gallery in one of the creeks running into Oratunga Creek, but had no idea what that meant or where it was. The local gallery is a place where carvings in stone were made thousands of years ago, and there are rare recorded sites in the Flinders Ranges. The locals, both the pastoralists and the Adnyamathanha people, aren't in the habit of being specific about the whereabouts of this place. There is a need to respect the fragility of the carved remains and the rights of the local Aboriginal community, and there is a healthy mistrust in the ability of visitors to look without touching, or more to the point, souveniring.

Upon entering the first creek, it's a process of negotiating the topography of the creek floor. In time, it is just possible to navigate your way across the boulders and outcrops without needing to stop at every new pattern or rock formation in the way that inveterate beachcombers stop to collect shells and the remains of sea creatures. Without a guide, the first-time trip is a little disconcerting, in the way that each distinct twist and turn seems to take on the sameness that is experienced by those who experience "bush blindness," with the concern that we are actually going around in circles.

After a particularly challenging navigation of creek bed the gallery rises up, and it is indeed a special space. There is unusually a sandy shore, and tufts of rushes are the evidence of a little spring just below the surface. In front, there rises up the most deeply textured wall that I have yet seen in this landscape. The myriad patterns are miniature topographies made over and over, over hundreds of thousands of years. The processes of fissuring, weathering, and crumbling enable multiple recordings concerned with naturally occurring surface and texture, some eerily reminiscent of a half remembered patterned interior or of a stylized freeway wall.

Rock topography: weathering/marking

I am enthralled by these surfaces, and this is indeed the gallery, where the imagery emerges from a vertical topography. The spatial sequencing of the Oratunga Creek walk shifts from the horizontal to the vertical along the organic pathway of the creek meander. The constant in each sequence lies in the variability of texture, where surfaces reveal the forces that have acted on them, layered over the underlying materiality of the geological strata. These creek beds are often smoothed, where intact slabs weather through water, rock moving against rock, sun and icy frost, heat and cold. Gorge walls shift through moments of compression, of rock fragments and sheets shearing off wall faces prised loose by tree roots or through water expanding to form ice chisels. Here and there, the ghosting of what appears as an imposed patterning emerges from under the weathered and textured rocks. Rough geometrical shapes—circles, arrows, and lines—can just be seen as the sun shadows form, etching both natural and human-made patterns more distinctly. And in the shadows, there looks like something that may be a lizard?

These scarred rocks reveal that this place may have been occupied much earlier than was previously known through the archeological evidence of found stone tools and occupation sites. It is thought that the Flinders Ranges may have been a "Pleistocene outpost" dating to about 15,000 years ago. This scarring is described in archeological terms as "the pecked and pounded designs of Aboriginal rock art," though the evidence is far from conclusive. If these marks are made by stone rubbing upon stone, then they have a date range between 1,000 and 30,000 years ago—hardly definitive, but these carvings may be associated with mythologies of the ancestral lizard. In other places, the elders have suggested that if the carvings are made by iron on stone, then they have been done by them (the men) recently and have no ceremonial significance. In any account, records suggest that the geometrical works are signs and maps that indicate where camps are to be found and give directions to other groups. They may be ancient signposts and records of things that once happened in everyday life in this fertile place, where the people gathered for water and shelter. These marks demand a respectful response to the wall; you sense that the correct action is one of not touching, merely pointing and wondering at the meaning of these simple scarrings. After the intensity of looking into the depths of the wall, a certain respite is needed; one steps backward to try to place this work within the spatial dimensions of this gorge, and within the context of this particular, brief journey into a landscape grounded by the unfamiliarity of age and slow yet indeterminate temporal change.

Some time after this first creek tour, I made a photographic montage based on surface works that had intrigued me. I was still enthralled by the walls in the Gorge, but now regard them as fragile places where discovery is possible only through guiding and paying close attention. So I made some works in disguise, through facile digital manipulations of my original photographic images, and presented them to others on a tabletop as a component of a collection of rocks and other debris collected from Oratunga; I sought to convey a visual and material mapping of the specifics of that walk in that landscape. I reproduce them to offer an intangible representation and a visual archeology of the Oratunga wallscape. In this image compilation, natural and reworked surface transforms into an abstracted three-dimensional map of the physical weathering and social marking of ancient landscapes, and it is a contemporary and ephemeral representation of a site-specific condition now displaced.

Through these walks, the practice of the intention to notice is honed through rewriting and re-presenting itineraries for the tour/s that I have made, which enables moving through arid landscapes that daily become more familiar. I also expand the itinerary by inviting others in to my moving about, facilitated by noticing unexpected means and chance associations. And they in turn speak of the tour with yet others, to sponsor subsequent movements later on.

NOTICING WHILE TOURING AND CURATING IN THREE WAYS: ON (RE)COLLECTION (RE)PRESENTATION IN THE HOUSE-GARDEN-MUSEUM

The third scenario draws on the archive and the museum. The discourses and practices of new museology seemingly embrace attitudes to collecting and archiving materials beyond classification and the repository of "valuable objects." Examining the collectors and curators who manage conceptually ephemeral material and house it in discursive space that is appropriated from some other use assists the curator to operate on the margins between conventional and experimental practices. Museums conceived through an ephemeral framing exist as sites of coincidence: between the space of the everyday here and now, and of the performative space of theater. Foucault writes in his fourth principle of heterotopia that museums operate within a temporal multiplicity because they endlessly accumulate yet ultimately enclose, conceptually and spatially. In Foucault's definition, the museum is a place that exhibits thematic yet often incompatible collections, framed through spatial itineraries that unfold over time.

I am often intrigued by the collections found in holiday houses and especially where the ephemera of the tourist, the walker, and the everyday explorer are carefully placed. The Oratunga Shearers' Quarters are usually devoid of decoration, but after every new visit things are left on the dresser or the windowsill, and the traces of one particular holiday are brought into the experience of the next. The instigators have no connection beyond occupying the same space at some time, but those who recognize the traces can often tell where others have been wandering. And one bird's nest or rock or discarded, rusting iron seems to prompt the next visitors to travel into the landscape and add more of the same. And then the arrangement and rearrangement of the collection follows, influenced through various curatorial and aesthetic preferences and negotiations supported by the vestiges brought in and left lying about. And over time a mini-archive emerges, a kind of material visitors' book; but often there are interruptions in the archive, because what is freely collected is also sometimes kept and transported elsewhere.

Through recording the abandoned Oratunga Blacksmith's Shed one morning, I demonstrate one approach to making the Oratunga archive resonant with vestiges of times past. Paraphrasing Foucault, I speculate whether from the fragment, over time, is it possible to represent the whole? The Blacksmith Shed is a collection inside a collection, impossible to capture spatially and materiality without compiling the fragments of the whole. I follow my touring method to spend time traveling around this small space and later reconstruct it in the studio to try to convey the whole to many others. As the montage develops, there is a curious doubling of certain things to be noticed: the light shards on the floor, the framed Medici man, old saddles, and a discarded shirt.

At Oratunga I work to develop the house-garden-museum through conceptualizing and cataloging the interstitial zones of the dry landscape, the collections of ephemera, and the event spaces in which the place is performed. Through curating the extant Oratunga homestead environs, relationships between archive and site, and the authentic and inauthentic, are identified and tested. I map and document the physical and temporal fabric of the homestead, including the events and incremental changes noticed and then participated in.

I am prompted to [re]think the structural concerns of the museum gallery space, although I am not really proposing the idea of the museum without walls for Oratunga; for that is an archive framed upon a scenographic aesthetic. The museum space I seek is a model where there are two conceptually defining elements upon which the museum hangs, recognizing that appropriated architectural space (house, yard, orchard, shed) is equally pertinent alongside the ephemeral museum, which performs through the small trace elements that make up this cultural landscape. The first element is the circulation space or spine that determines movement, yet in the ephemeral museum, the spine leads and then drops one in space or place, demanding only that venturing and returning is an exploratory act. The second element is the conceptual net thrown over the museum space: it contains the abstract volume, frames connections between the interior and the exterior (physically and culturally), and gives form to the collections and performances housed within.[10] These elements remain the conceptual challenge for conceiving the gallery at Oratunga. In many ways, though, the greater experiment is to curate the means to represent and translate the multiplicity of landscape and event qualities that create and craft the physical and temporal spaces of Oratunga.

ON PERFORMATIVE DESERT/ARID TOURISM

Remote desert places are the repositories of a rich array of visual references on which design practice often depends for source material for the tourist market. Yet these places are also fragile environments ecologically and culturally, and although they require

from the visitor robust intervention to weather arid conditions, the fabric of their histories disappears in a moment through inattention and/or an inability to respond appropriately. This research suggests ways of noticing through curating landscapes and places that inform design practices concerned to build on ephemeral conditions.

I have described three scenarios that situate subtle curatorial design practices in arid environments to invite a more performative interaction in marginal and often difficult places. The essential attribute of these examples is that they acknowledge that the present condition reveals past traces and occupations, in turn to postproduce the present. I see a curatorial approach that implies rather than prescribes design outcomes as pertinent to working in desert places, but not simply because they are usually subject to scant resources and expertise for making new architectures. Desert tourism in arid Australia needs to span and mediate the practices and discourses of extant built fabric, local cultures, and the global tourist. A curatorial attitude enriches such tourism through proposing interventions that enable the translation of desert knowledge to new and unfamiliar audiences.

NOTES

1   Virginia Lee, "On the Sublime: A Regard for Landscapes and Ruins, Yet Not Really about the Picturesque," in *The Intention to Notice,* doctoral durable visual record, RMIT University, 2006, 91–92.

2   Nicolas Bourriaud, *Postproduction: Culture as Screenplay: How Art Reprograms the World* (New York: Lukas & Sternberg, 2002). Postproduction theory and practice as applied to landscape design informs the theoretical framework for this work.

3   The Placecards Project was supported by the Centenary of Federation 2001, South Australia.

4   Italo Calvino, *Six Memos for the Next Millennium* (Cambridge, MA: Harvard University Press, 1988), 16.

5   Ibid., 3.

6   John Rajchman, *Constructions* (Boston: MIT Press, 1998), 39.

7   Ibid., 51.

8   Claire Doherty, "The New Situationists," in Claire Doherty, ed., *Contemporary Art: From Studio to Situation* (London: Black Dog Publishing, 2004), 11

9   Robert Smithson, "A Provisional Theory of Non-Sites" (1968), in Jack Flam, ed., *Robert Smithson: The Collected Writings,* 2d ed. (Berkeley: University of California Press, 1996), 364.

10 I acknowledge appropriating Colin Fournier's description of his museum strategies, which I heard at an RMIT University lecture in 2004.

Contributors

Vincent Battesti is an anthropologist at the Museum National d"Histoire Naturelle, Paris. A social anthropologist, he is interested in the relationships between communities and their environments in the oases of the Sahara and in urbanscapes (Cairo, Khartoum). He has worked on economic development (Tunisia, Vanuatu) and in university-based (Yemen) and other research centers (Paris, Cairo).

Aziza Chaouni (BSc, MArch) is a structural engineer and architect. She has researched desert tourism in the Sahara under the Appleton Traveling Fellowship, awarded by the Harvard Graduate School of Design. She was an Aga Khan Visiting Fellow as well as a Visiting Design Critic in Urban Planning and Design. She teaches in the John H. Daniels Faculty of Architecture, Landscape, and Design at the University of Toronto.

Chris Johnson (BSc, MSc) is an ecologist and conservationist, specializing in the institutional strengthening of environmental NGOs and the development of ecotourism and other nature-related businesses. He started working in Jordan in 1994 to help establish the Dana Nature Reserve as a model of integrated conservation and development. He is now the Director of "Wild Jordan," the division responsible for ecotourism and socio-economic development at the Royal Society for the Conservation of Nature.

Gini Lee is an interior designer and landscape architect who teaches, researches, and practices in multidisciplinary design areas including domestic and commercial interiors, museum and exhibition design, environmental graphics, garden design, indigenous housing and landscape design methodology, and cultural and historical landscapes" assessment and interpretation. From 1999 to 2003, she was Head of School at the Louis Laybourne Smith School of Architecture and Design.

Neil Levine teaches the history of modern architecture at Harvard University, where he is the Emmet Blakeney Gleason Professor of History of Art and Architecture. He has been the Banister Fletcher Professor of Architecture at the University of London and the Slade Professor of Fine Art at Cambridge University. He is on the editorial board of the journal *Wright Studies* and was on the advisory board of a PBS documentary series on Frank Lloyd Wright. He is also on the Board of Directors of the Frank Lloyd Wright Building Conservancy.

Susan Gilson Miller is Associate Professor of History at the University of California, Davis. Prior to that, she was Director of the Moroccan Studies Program at Harvard University and Senior Lecturer in the Department of Near Eastern Languages and Civilizations. She teaches courses on Maghribi social and cultural history and Mediterranean urbanism. Her primary area of research is North African urban history in the colonial period. She is the editor, with Mauro Bertagnin, of *The Architecture and Memory of the Minority Quarter in the Muslim Mediterranean City* (2010).

Virginie Picon-Lefebvre is an architect and urban designer. She is a Professor at the Ecole Nationale Supérieure d"Architecture Paris-Malaquais and was a lecturer at the Harvard Graduate School of Design during 2001–2007. She has a PhD in Architectural History and teaches courses in urban design. She has been Professor of Architecture and Urban Design at the Ecole d"Architecture in Versailles, France, and a visiting scholar at MIT. She is the author of *La Grande Arche de la Defense* (1989) with Francois Chaslin, and the editor of *Les espaces publics modernes* (1995). In 1988, she founded a research group with Claude Prelorenzo, GRAI (now LIAT), to study the relationship between architecture and infrastructure. She is curently working on a book about the architecture of tourism.

Alessandra Ponte teaches at the École d"Architecture, Université de Montréal. She was Associate Professor at the School of Architecture, Pratt Institute, and has also taught at Princeton, Cornell, and the ETH in Zurich. She has published a book on Richard Payne Knight and the picturesque (2000) and coedited, with Antoine Picon, a collection of essays on architecture and the sciences (2003). She is currently preparing a book on the American desert.

Claude Prelorenzo is a sociologist and urban planner. He was the director of an international research group, GRAI (now LIAT), founded with Virginie Picon-Lefebvre in 1988 to study the relationship between architecture and infrastructure. He is the general secretary of the Fondation Le Corbusier. He has directed the program of research Le Port et la Ville and published several books on this subject. He is curently teaching in the masters program for urban planning at the Ecole Nationale des Ponts et Chaussees.

Robert Sumrell is Adjunct Assistant Professor at the Columbia University Graduate School of Architecture, Planning, and Preservation. He began working in set design after nearly a decade of practicing in interiors and architecture firms, where he worked on projects ranging from high-end residences, retail stores, sports stadiums, dot.com offices, urban planning, and charter schools. As a cofounder of the nonprofit research group AUDC (with Kazys Varnelis), his work has been curated in gallery shows and published in numerous books and periodicals.

Kazys Varnelis is Director of the Network Architecture Lab and teaches studio and history-theory at the Columbia University Graduate School of Architecture, Planning, and Preservation. He is also a founding partner of the speculative architectural research group AUDC. He has published widely in journals such as *Praxis, Log,* and *Verb.* He is the cofounder, with Robert Sumrell, of AUDC, and coauthor of *Blue Monday* (2007). His areas of research include contemporary architecture, late modernism, architecture and capitalism, and the impact of recent changes in telecommunications and demographics on the contemporary city.

Acknowledgments

This book, and the conference and seminar that informed its content, would not have been possible without the help and support of a large group of people at Harvard University at large, the Aga Khan Program at the Harvard University Graduate School of Design, the Moroccan Studies Center at Harvard University, and the Moroccan Ministry of Tourism.

Beginning in 2005, the Aga Khan Program at the GSD encouraged the exploration of research approaches and theoretical frameworks with which to tackle the development of tourism in desert landscapes. Also, the year 2006 was designated by UNESCO as the year of the desert, and the idea of the link between development and sustainable tourism was promoted.

In fall 2006, the GSD initiated a collaboration with the Moroccan Minister of Tourism, Adil Douiri, on the development of tourism in the fragile desert regions of southern Morocco. A research seminar and an international conference on the theme of desert tourism were organized by Virginie Lefebvre and Aziza Chaouni during spring 2007, with the support of the Chair of the Urban Design Department, Professor Rodolfo Machado, as well as Executive Dean Pat Roberts. The seminar involved a trip to the desert city of Merzouga. Omar Bennani, Moha Errich, and Abdelkhalek Chkoukout from the Ministry of Tourism were instrumental in organizing the trip and facilitating field research. The students in the seminar worked hard to collect information and designed proposals for the reconstruction of the village.

The research work and discussions that emerged from the seminar helped us frame the challenges, impacts, and cross-disciplinary potentials inherent to desert tourism. The student research from the seminar uncovered varied forms of tourism in deserts worldwide. Professor Mauro Bertagnin was very kind to introduce our students to rammed earth construction techniques and share with us his experience in Mali. Elise Newman gave an important lecture about cartography. The conference was supported by a Harvard University Provost Grant, awarded to Virginie Lefebvre and Aziza Chaouni. Professors Susan Miller, then Director of the Moroccan Studies Center, Hashim Sarkis, Neil Levine, and Steven Caton were instrumental in helping us frame the conference. While preparing for the conference, volunteer students" and GSD staff"s help was crucial. We would like to especially thank Nayla Al Akl, Roula El Khoury, Joshua Haddad, and Maria Moran.

We would also like to thank Ann Pendelton-Jullian, Michael Meredith, and Robert France for introducing speakers during the conference, and Ali Amahan, Fares Alswaidi, Alexia Leon, Steven Caton, Moha Errich, Rick Joy, and Ashraf Salama for their lectures; as well as Vincent Battesti, Chris Johnson, Neil Levine, Gini Lee, Susan Miller, Alessandra Ponte, Claude Prelorenzo, Robert Sumrell, and Kazys Varnelis for their collaboration during the long process of editing this book.

Our warmest thanks go to Professor Hashim Sarkis, who offered unwavering support for this project and enabled the collaboration within the GSD of Professor Virginie Lefebvre and Aziza Chaouni, then an Aga Khan Fellow at the GSD.

Illustration Credits

8, 11, 15 (bottom) Normal F. Carver, Jr., *North African Villages* (Kalamazoo, Michigan: Documan Press, 1989), pp. 41, 82.

10 ©2011 Google, map data ©2011 Europa Technologies, Google, Tele Atlas.

13 Charles de Foucauld, *Reconnaissance au Maroc: Journal de route* (Paris: Société d'editions, 1939).

14 Abdelaziz Ghozzi, *The Orientalist Poster: A Century of Advertising through the Slaoui Foundation Collection* (Casablanca, Morocco: Malika, 1997), p. 48.

15 (top) D. Jacques-Meunié, *Architectures et habitats du Dadès, Maroc présaharien* (Paris: C. Klincksieck, 1962).

16–17 Walter B. Harris, *Tafilet: The Narrative of a Journey of Exploration in the Atlas Mountains and the Oases of the North-west Sahara* (Edinburgh: W. Blackwood and Sons, 1895), p. 277.

18 Larbi Mezzine, *Le Tafilalt: Contribution à l'histoire du Maroc aux XVIIe et XVIIIe siècles* (Rabat: Faculté des lettres et des sciences humaines, 1987), p. 135.

20, 31 (bottom) "Lawrence of Arabia" ©1962, renewed 1990 Columbia Pictures Industries Inc. All rights reserved. Courtesy of Columbia Pictures. Peter O'Toole image used with permission Steve Kenis and Company.

25 By permission of The Imperial War Museum, image Q73535.

31 (top) Paramount Pictures.

36 (top) Ralph Crane © Black Star.

36 (bottom) Kate Chesley.

38 (top) National Park Service, Scotty's Castle, Death Valley, California.

34, 38 (bottom), 42 (top), 42 (bottom), 43, 44, 45 The Frank Lloyd Wright Foundation, Scottsdale, Arizona.

40 From Federal Writers' Project of the Works Progress Administration of Northern California, *Death Valley: A Guide*, 1939.

41 Courtesy Arizona Biltmore Hotel.

46 Herb McLaughlin. Courtesy: Dorothy McLaughlin, Arizona Photographic Associates.

47 Private collection.

48 Ezra Stoller © Esto.

55 (top) Private collection of Faraoui and Demazières and photos by the author.

56 (left) Moroccan Ministry of Tourism.

56 (right), 58 Companie Générale Transatlantique.

59, 60 (top) Private collection of Faraoui and Demazières.

All other images courtesy of the authors.

The Aga Khan Program for Islamic Architecture at Harvard and MIT

Based at Harvard University and the Massachusetts Institute of Technology, the Aga Khan Program for Islamic Architecture (AKPIA) is dedicated to the study of Islamic art and architecture, urbanism, landscape design, and conservation, and the application of that knowledge to contemporary design projects.

The goals of the program are to improve the teaching of Islamic art and architecture, promote excellence in advanced research, enhance the understanding of Islamic architecture, urbanism, and visual culture in light of contemporary theoretical, historical, critical, and developmental issues, and increase the visibility of Islamic cultural heritage in the modern Muslim world. Established in 1979, AKPIA is supported by an endowment from His Highness the Aga Khan. AKPIA's faculty, students, and alumni have played a substantial role in advancing the practice, analysis, and understanding of Islamic architecture as a discipline and cultural force.

The Aga Khan Program at the Harvard University Graduate School of Design

Established in 2003, the main aim of the Aga Khan Program at the GSD is to study the impact of development on the shaping of landscapes, cities, and regional territories in the Muslim world and to generate the means by which design at this scale could be improved.

The program focuses on the emerging phenomena that characterize these settings and on issues related to the design of public spaces and landscapes, environmental concerns, and land use and territorial settlement patterns. The process entails a study of their current conditions, their recent history (from World War II to the present), and, most important, the exploration of appropriate design approaches. The program sponsors new courses, option studios, faculty research, workshops, conferences, student activities, and publications. It is supported by a generous grant from the Aga Khan Trust for Culture.

TITLES IN THE AGA KHAN PROGRAM BOOK SERIES

Two Squares: Martyrs Square, Beirut, and Sirkeci Square, Istanbul,
edited by Hashim Sarkis, with Mark Dwyer and Pars Kibarer

A Turkish Triangle: Ankara, Istanbul, and Izmir at the Gates of Europe,
edited by Hashim Sarkis, with Neyran Turan

Han Tumertekin: Recent Work, edited by Hashim Sarkis

The Architecture and Memory of the Minority Quarter in the Muslim Mediterranean
City, edited by Susan Gilson Miller and Mauro Bertagnin

The Superlative City: Dubai and the Urban Condition in the Early Twenty-First
Century, edited by Ahmed Kanna

Landscapes of Development: The Impact of Modernization Discourses
on the Physical Environment of the Eastern Mediterranean, edited by Panayiota Pyla